Montréal at the Crossroads

# Montréal at the Crossroads
## Superhighways, the Turcot and the Environment

*Edited by*
Pierre Gauthier
Jochen Jaeger
Jason Prince

**BLACK
ROSE
BOOKS**

Montreal · New York · London

Black Rose Books No. NN372

Catalogage avant publication de Bibliothèque et Archives Canada

Montréal at the crossroads : superhighways, Turcot and the environment / edited by Pierre Gauthier, Jochen Jaeger and Jason Prince.

Textes en anglais et en français.
Comprend des références bibliographiques.
ISBN 978 1 55164 342 7

1. Urbanisme—Québec (Province)—Montréal.  2. Transport—Québec (Province)—Montréal.  3. Environnement—Québec (Province)—Montréal. I. Gauthier, Pierre, 1963–  II. Jaeger, Jochen A. G  III. Prince, Jason, 1965–

HT169.C22M655 2009          307.76'40971428          C2009-903070-5F


C.P. 1258                    2250 Military Road          99 Wallis Road
Succ. Place du Parc          Tonawanda, NY               London, E9 5LN
Montréal, H2X 4A7            14150                       England
Canada                      USA                         UK

To order books:
In Canada: (phone) 1 800 565 9523 (fax) 1 800 221 9985
email: utpbooks@utpress.utoronto.ca
In the United States: (phone) 1 800 283 3572 (fax) 1 800 351 5073
In the UK & Europe: (phone) 44 (0)20 8986 4854 (fax) 44 (0)20 8533 5821
email: order@centralbooks.com
Our Web Site address: http://www.blackrosebooks.net

Book designed and readied for the press by Zack Taylor.

Front cover design and photographs by David Chedore.

Printed in Canada at Gauvin Press.

**Mixed Sources**
Product group from well-managed forests and other controlled sources
www.fsc.org Cert no. SGS-COC-2624
© 1996 Forest Stewardship Council
FSC

# Contents

Colour plates are found in the middle of the book.

# Wrong Road Forward: The New Turcot Project Fails to do Much to Accommodate Trains and Other Public Transit[1]

*Henry Aubin*
*Montreal Gazette, Saturday, March 28, 2009, page B7*

The provincial government's plan to remake the Turcot Interchange is drawing criticism from a number of urban planners, environmentalists and nearby residents.

These critics don't contest the Department of Transport's argument that the existing aerial spaghetti of roads is in danger of collapsing and that quick corrective action is imperative. But they object to the ministry's plan to bring the roadways down to ground level where they could accommodate 310,000 cars a day, a 10-per-cent increase over the current level.

Which side is right?

An engineering report last year confirmed that the elevated roads are dangerously decrepit. The ministry's $1.5-billion plan to bring them to ground level by 2016 would be cheaper than rebuilding them in the air.

But the critics' points are also strong. They point out the plan ignores the fight against climate change. And although the plan calls for reserved bus lanes, its thrust would be to perpetuate the supremacy of commuting by car. Motor vehicles generate about a third of Quebec's greenhouse-gas emissions as well as much of the region's smog.

Another problem is that the grassy embankments on which the roads would be located would have broad shoulders and be about three times as wide as today's skinny aerial roads (call it the fettuccini effect). Today, cyclists and pedestrians can pass freely under the elevated roads; the embankments would prevent such coming and going, cutting off parts of the city from each other. (The occasional tunnels for cyclists and pedestrians would likely pose security problems).

As well, the new configuration would require the expropriation of 160 housing units. The roads' location at ground level would also increase the noise and air pollution for the remaining nearby residents.

Several urban planners have offered solutions to these neigbourhood issues. One scenario would be to build all or most of the interchange underground. Costs, however, would soar: The ministry says that tunnels would require pumping stations and ventilation. Trucks carrying dangerous goods could not enter.

Another scenario would be to prop the existing elevated structure with metal arches, but you have to wonder how permanent a solution that would be.

But let's get back to the climate-change problem. The Obama administration and many experts say that a future international treaty must aim for an 80-per-cent (at least) reduction in global greenhouse-gas emissions by 2050. Oblivious, Quebec's transport department is building the wrong infrastructure for compliance.

Mayor Gérald Tremblay urged Transport Minister Julie Boulet last fall to make more room on the new interchange for public transit, but we haven't heard much from him since. In the case of the Highway 25 extension and the Notre Dame St. enlargement, Tremblay showed a flair for initially chiding the department, then caving.

Tremblay also asked Boulet that the new interchange be designed in such a way as to allow room for a future shuttle train between downtown and Trudeau International Airport. If such a frequently running rail line were to continue on to Ste. Anne de Bellevue, as many people hope, it could carry a great number of the West Island commuters who now drive through the interchange.

The problem is not in adapting the interchange to the train. It's in getting the train. The idea has been in the wind since the 1996 decision to transfer most flights from Mirabel to what is now Trudeau, but talks between Quebec and its partner in the concept, Ottawa, have produced nothing. Thirteen years of dithering.

If such a commuter train and other major public-transit schemes were in the works, the ministry might be able to scale down its interchange plan significantly.

The longer the provincial and federal governments delay in shifting transport policy away from cars and toward less environmentally harmful substitutes, the longer we'll be stuck with costly interim projects like the Turcot Interchange. It's a dumb project, but the public needs a safe alternative to the present ruin, and no other alternative is in sight.

### Note

1. Reprinted with permission from the *Montreal Gazette*.

# Paris or Detroit?

*John Norquist*

The Turcot Interchange is a blight that needs to be removed. Enlightened cities around the world are removing freeways and replacing them with surface streets. Seoul, South Korea, San Francisco, Stockholm, New York and dozens of other places are removing grade separated roads and restoring streets that add value and distribute traffic better especially at peak times.

Tearing down freeways may seem crazy or at least counterintuitive, but Montrealers should ask themselves what model they prefer for their city, Paris or Detroit. Paris has no freeways in the city; Detroit is covered with grade separated roads. Paris is complex, congested and crowded; Detroit is almost empty. Traffic congestion has been defeated in Detroit. It is still a problem in Montreal and Paris. Montreal can defeat congestion or be a great city. It's your choice.

# Montréal à la croisée des chemins

Au moment où cet ouvrage est mis sous presse, Montréal est plongée dans un débat au sujet de l'échangeur Turcot.

Le plus grand échangeur de ce type au Québec, ce carrefour autoroutier aérien, permet quotidiennement à plus de 280 000 véhicules—automobiles, camions, autobus et autocars—d'atteindre l'un ou l'autre des quatre coins de la ville. L'échangeur Turcot fut construit dans les années 1960, juste à temps pour l'Exposition universelle de Montréal. Maintenant âgée de 40 ans, la structure se fait vieille. Que doit-on faire ?

Pendant des années, les ingénieurs du ministère des Transports du Québec (MTQ) se sont vaillamment ingéniés à développer une proposition qui remplacerait cette infrastructure vétuste. Leur solution suggère de substituer, à l'ancienne structure aérienne, une nouvelle, construite sur talus.

Il importe de préciser que la proposition du MTQ ne touche pas uniquement le nœud enchevêtré de l'échangeur Turcot qui raccorde les autoroutes 15 (dans l'axe nord-sud) et 720 (dans l'axe est-ouest), mais qui relie également deux autres carrefours autoroutiers situés respectivement au sud et à l'ouest (les échangeurs de la Vérendrye et Montréal-Ouest) ainsi que des kilomètres d'autoroutes attenantes, déployées tantôt en surface, tantôt dans les airs (*cf.* les illustrations couleur au milieu de cet ouvrage).

En tablant sur un accroissement modeste de la capacité routière, le MTQ prévoit une augmentation du volume de circulation ; le nombre de véhicules utilisant l'échangeur à proprement parler et ses portions autoroutières contiguës passerait de 332 000 (en 2009) à 347 000 par jour (en 2016). Le Ministère ose croire qu'une telle hausse ne se traduira pas par une augmentation de la pollution ambiante, en misant sur une efficacité accrue des véhicules automobiles en matière d'émissions polluantes. Bref, plus d'automobiles, mais moins de pollution ! Nulle mention des mesures à prendre pour que le nombre de kilomètres parcourus n'augmente pas cependant.

Un consensus semble néanmoins se développer chez les critiques du projet à l'égard d'un aspect important : quoiqu'il advienne de l'échangeur, l'axe autoroutier nord-sud et son volume de circulation devront probablement être conservés du fait qu'il n'y a pas actuellement d'alternatives viables pour remplacer ce segment critique de notre système autoroutier national. La pertinence de la reconstruction du corridor est-ouest parallèlement à l'axe de l'autoroute 720 sous une forme autoroutière est cependant beaucoup plus discutable. Cet axe sert présentement une majorité de navetteurs qui pourraient fort bien emprunter les transports en commun. De fait, ceux-là constituent précisément la clientèle-cible des usagers potentiels visés par les villes qui prennent des initiatives touchant à la fois le démantèlement des autoroutes urbaines polluantes et intrusives et la bonification des systèmes de transport en commun.

On argue que le maintien et l'entretien de la nouvelle structure autoroutière seront plus aisés et moins onéreux, allégeant ainsi le fardeau financier transmis aux générations futures (ce qui correspond à la façon dont le MTQ se représente le « développement soutenable »). Le nouveau carrefour autoroutier sera de même construit, nous dit-on, selon les standards autoroutiers modernes, avec accotements, destinés à accroître la sécurité des usagers.

Les opposants au projet font valoir qu'une telle perspective est inadéquate à plusieurs égards. La crise des changements climatiques ainsi que les problèmes de santé associés à notre dépendance collective à l'automobile commandent plus de discernement et appellent des actions plus énergiques. D'abord, partout où faire se peut, des modes de transport plus soutenables doivent être offerts afin d'offrir une autre « voie » que l'automobile. Il faut reconnaître ensuite que les quartiers centraux densément peuplés, qui furent mis à mal par la construction des autoroutes urbaines dans les années 1960 et 1970, sont ceux-là mêmes où il convient aujourd'hui de canaliser le développement. Cela vaut en particulier pour les anciens sites de production, aujourd'hui en jachère, qui truffent ces quartiers. Dans sa forme actuelle, non seulement la proposition du MTQ n'offre-t-elle rien pour réparer les erreurs du passé en investissant dans l'amélioration de la qualité de vie à Montréal, mais elle aurait pour effet de sacrifier, et ce pour plusieurs générations, le potentiel de développement d'une portion vitale du cœur de la cité.

De plus, les opposants font valoir que la construction d'une autoroute sur talus équivaut à appliquer une solution rurale à un problème urbain. Des talus n'ont pas leur place en ville : la proposition du MTQ produit des murs qui divisent les quartiers, des culs-de-sac ainsi que des fragments et des espaces résiduels à peine utilisables et marqués par une valeur dévaluée qui affecte négativement l'assiette foncière de la ville. La construction sur talus implique en outre le triplement de l'emprise au sol de l'ouvrage, qui se traduit par un gaspillage accru de précieux hectares des quartiers centraux.

Selon la proposition du MTQ, la nouvelle chaussée doit être construite en parallèle de la chaussée existante afin de ne point interrompre le flot de circulation pendant le cours des travaux. À terme, la circulation sera redirigée vers la première, permettant ainsi la démolition de l'infrastructure d'origine. L'espace additionnel requis par ce scénario de construction implique l'expropriation et la démolition de quelque 200 logements de bonne qualité, situées dans le village des Tanneries, une petite enclave du quartier Saint-Henri (les automobilistes qui empruntent les rampes d'accès vers l'est depuis l'autoroute Décarie peuvent jeter un regard furtif sur la fenêtre ornée de rideaux d'un loft qui se trouve à moins de 10 mètres de la courbe bétonnée ; c'est dire la proximité de certaines habitations !).

Le Ministère a offert de faire reconstruire « tout près » des unités d'habitation de remplacement abordables, et de compenser les propriétaires et les locataires en accord avec les exigences de la loi québécoise. Mais plusieurs résidents risquent de ne pas avoir droit à ces logements en raison des critères restrictifs des programmes en place. Les résidents sont outrés à juste titre de la perspective de la démolition d'une partie de leur quartier et craignent que ce qui en restera ne survive pas au traumatisme.

Le MTQ promet de pourvoir les talus d'une dense végétation, mais les résidents demeurent sceptiques, car la revue des autoroutes québécoises ne produit rien de comparable aux rendus d'artistes produits par le Ministère à l'occasion des consultations publiques.

En 2006, le MTQ estimait le coût de son projet de construction sur talus à 1,5 milliards de dollars. Il est difficile de prendre une telle estimation des coûts au sérieux, pour des raisons que les Montréalais n'ont pas de peine à comprendre.

Un exemple récent d'escalade des coûts devrait suffire à rappeler le point. Le coût estimé du projet de construction de l'autoroute Notre-Dame, à l'est du centre-ville de Montréal, est passé, en moins de deux ans, de 750,000$ (le coût projeté en 2006) à 1,5 milliards de dollars aujourd'hui. Le MTQ s'est empressé d'y mettre le holà et en est à revoir le dossier. La Ministre avance qu'un tel dépassement de coût n'est pas à prévoir dans le cadre du projet Turcot, où l'on a pris soin d'inscrire des imprévus au budget ; mais plusieurs personnes demeurent sceptiques.

★ ★ ★

Chaque jour, des centaines de milliers de personnes empruntent l'échangeur Turcot. D'autres personnes, par centaines de milliers, entendent parler du sort des premières à la radio du matin et à celle du « retour à la maison ». Certaines personnes—appelons-les les non-automobilistes—, se représentent l'échangeur comme un lieu plutôt mystérieux, qu'ils imaginent débordant d'automobiles, tantôt inanimées parce qu'interrompues dans leurs déplacements, tantôt se déplaçant avec une enviable fluidité. Pour d'autres personnes encore, quittant chaque matin la maison pour le travail au volant de leur voiture, l'échangeur est un passage quotidien quasiment obligé, et un coup de dés : « Serai-je embourbé dans la circulation encore aujourd'hui ? » Ces Montréalais se meuvent l'oreille collée à la radio, supputant la meilleure façon de naviguer dans les rues de la ville jusqu'à destination.

&#9733;  &#9733;  &#9733;

Des actions sont requises à l'égard de l'échangeur Turcot, tous en conviendront. Le MTQ prétend avoir envisagé la rénovation de l'ouvrage en 2004, avant de rejeter une telle éventualité parce que trop coûteuse et requérant plus de temps que le scénario de « démolition-reconstruction » pour lequel il a finalement opté. Mais les documents qui témoignent de cette démarche sont demeurés à usage interne et ne furent pas soumis à la consultation du public.

Depuis l'effondrement, en 2006, du viaduc du boulevard Concorde, tragédie qui a enlevé la vie à cinq personnes en plus d'en blesser d'autres, le gouvernement du Québec s'est voulu très vigilant. Un article  paru dans le quotidien Le Devoir, en octobre 2008, disait que l'échangeur Turcot était « la structure la plus surveillée au Québec ». Des centaines de réparations ont été effectuées sur la structure au cours des cinq dernières années. Les automobilistes croisant régulièrement, sous les immenses rubans de béton de l'échangeur, les treillis d'armatures métalliques installés sous les structures par mesure de sécurité, ne peuvent réfréner le réflexe de leur jeter un regard nerveux.

Plusieurs automobilistes anticipent en effet un effondrement, mais le monitorage de tous les instants effectué par les spécialistes aide à inspirer une certaine confiance. La population continue à utiliser l'ouvrage.

L'échangeur Turcot peut-il être réparé ? Sa capacité autoroutière doit-elle être réduite tout en assurant la mobilité des personnes par d'autres moyens tels les trains, les tramways et les autobus ? Devrait-on simplement démanteler le tout ? Doit-on au contraire sauvegarder l'ouvrage à titre de monument et de témoignage d'une urbanité révolue ?

&#9733;  &#9733;  &#9733;

Le présent ouvrage se veut une importante contribution aux débats sur le sort de l'échangeur.

Il est crucial de tenir un tel débat immédiatement. Le MTQ souhaite entreprendre des travaux dès la fin de 2009. Les audiences publiques tenues sous l'égide du Bureau d'audiences publiques sur l'environnement (BAPE) ont commencé promptement (en vertu d'un processus accéléré mis en place par un gouvernement impatient de voir le projet se mettre en branle). D'une part, les résidents, les groupes communautaires et les experts eux-mêmes sont enterrés sous la masse d'informations produites en marge de l'étude d'impact environnemental—étude qui est, à strictement parler, le seul objet des audiences publiques. D'autre part, ces résidents, ces groupes et ces experts déplorent le manque d'informations cruciales dont ils ont besoin pour se faire une opinion bien informée.

Le Conseil régional de l'environnement de Montréal, un groupe d'influence important, a demandé—et a, en partie, obtenu—la tenue d'un débat élargi à la problématique d'ensemble des transports dans l'axe est-ouest de la métropole dans le cadre des audiences du BAPE. La lettre que le Conseil a soumise à Mme Line Beauchamp, ministre du Développement durable, de l'Environnement et des Parcs est reproduite à l'annexe B. Un second groupe a entrepris des démarches légales appelant à la suspension et au report des audiences jusqu'à ce que de l'information additionnelle soit rendue publique. Un groupe constitué d'urbanistes et de professeurs de même que de différents groupes communautaires argue que la solution mise de l'avant par le MTQ est une réponse technicienne renvoyant à une mauvaise façon de poser la question (en l'espèce : comment assurer le déplacement de tel volume de circulation, à telle vitesse, vers telles destinations ?). Le débat devrait plutôt porter sur la problématique d'ensemble et sur la réponse globale à apporter en rapport avec des objectifs de mobilité des personnes dans une ville désormais mieux conçue. Les représentations faites par ce groupe sont reproduites à annexe A.

Dans le monde, bien des villes et leurs résidents se demandent comment régler le problème des infrastructures vieillissantes construites pour l'automobile durant l'ère du pétrole à bon marché. Montréal ne fait pas exception et ne peut se dérober à ce débat.

Le présent ouvrage propose un éventail de points de vue qui, espérons-nous, enrichira la discussion et suscitera de nouvelles manières de penser le futur de notre ville.

Nous souhaitons que cette modeste contribution stimule les débats et contribue à l'avènement d'un Montréal amélioré, pour le bénéfice de l'ensemble de ses citoyens.

★　★　★

Cet ouvrage est organisé en fonction des grands thèmes qui ont marqué les débats de la dernière année dans des dizaines de cercles de gens et de groupes concernés à travers la ville.

Certaines des contributions y sont le fait d'experts et de chercheurs reconnus, d'autres sont le fruit du travail de réflexion mené par des étudiants sur des sujets précis et d'autres encore nous parviennent de membres de la nouvelle génération d'activistes.

Ce livre présente de la sorte un éventail de voix. Les paragraphes qui suivent entendent guider le lecteur à travers les différents chapitres et les différents thèmes de l'ouvrage.

Le système de transports de l'agglomération montréalaise est caractérisé à la fois par une offre substantielle de transports en commun *et* par une dépendance à l'égard de l'automobile. Dans le chapitre 1, **Jeff Kenworthy** et **Craig Townsend** affirment que la reconstruction de l'échangeur Turcot, telle que proposée, sans égard au développement du transport en commun, aurait pour effet d'accroître la dépendance à l'automobile et d'affaiblir la compétitivité des réseaux de transports collectifs. Les auteurs présentent des données qui comparent le système de transports de Montréal à celui d'autres villes du monde pour en conclure que la conjoncture montréalaise appelle une amélioration de son transport en commun et ne requiert pas une augmentation du nombre de voies routières à grand débit.

Le chapitre 2 présente des contre-propositions au scénario soumis par le ministère des Transports du Québec (MTQ). **Pierre Brisset** et **Jonathan Moorman** plaident en faveur d'une approche centrée sur le transport en commun qui soit en accord avec le Plan de Transport de la Ville de Montréal (2008) pour le re-développement du complexe de l'échangeur. Leurs propositions associent une bonification de l'offre de transport en commun et des mesures astreignantes pour les automobilistes en vue d'une réduction du nombre de ceux d'entre ces derniers qui font la navette quotidienne entre l'ouest de l'île ou les quartiers centraux du Sud-Ouest de Montréal et le centre-ville. De telles mesures permettraient de revoir la conception de l'échangeur autoroutier et celle de l'autoroute Ville-Marie en vue de réduire leur taille en trois phases subséquentes. Les auteurs proposent une vision stimulante selon laquelle, à terme, l'autoroute Ville-Marie disparaîtrait et l'échangeur serait converti en carrefour à trois embranchements.

Le chapitre de **Pieter Sijpkes** gratifie cet ouvrage d'un plaidoyer vibrant et éclairé. Dans le chapitre 3, Sijpkes soutient que la proposition du Ministère pour l'échangeur Turcot est malavisée à deux égards. D'abord, plutôt que de démolir la structure de béton, nous devrions focaliser sur ce qui est toujours viable. Les sections aériennes de l'autoroute pourraient ainsi être supportées par des poutres d'acier. Cette approche aurait pour effet, avec un coût réduit, de prolonger la durée de vie utile de l'ouvrage de plusieurs années, tout en facilitant les inspections et la maintenance, et ce—argument crucial—sans interrompre la circulation pendant les travaux ni exiger la démolition de résidences. Ensuite, plutôt que de déplacer l'autoroute 720 et les voies ferroviaires de la compagnie des chemins de fer du CN sur le piedmont de la falaise Saint-Jacques, Sijpkes suggère d'agrandir le parc linéaire de la falaise sur son piedmont, de restaurer la rivière Saint-Pierre depuis longtemps asséchée et de concevoir avec soin un écran végétal destiné à isoler le parc de l'autoroute.

**Patrick Asch** pousse plus avant cette idée de construction d'un grand parc urbain au chapitre 4, en avançant l'argument selon lequel la falaise Saint-Jacques est un joyau insoupçonné du réseau montréalais d'espaces verts. Le parc de la falaise rivalise en superficie avec le parc du Mont-Royal et le fait d'y adjoindre quelques liens verts stratégiquement situés dans des emprises existantes ferait de ce parc linéaire une armature verte connectant plus de 500 hectares d'espaces verts dans les secteurs sud et ouest de l'île de Montréal. Asch avance également l'idée audacieuse de troquer des hectares de terrain du site Meadowbrook, dont le sort est l'objet de vives discussions, contre une superficie équivalente dans l'ancienne cour de triage Turcot afin de permettre le développement résidentiel dans le premier plutôt que dans la seconde. Asch nous instruit du fait que dans sa forme actuelle, la proposition du MTQ de déplacer l'autoroute au pied de la falaise aurait pour effet de détruire cet espace vert en sacrifiant les fonctions naturelle, sociale, voire économique, qu'il est appelé à servir.

Dans le chapitre 5, **Raphaël Fischler** nous redit à quel point l'aménagement du territoire est compliqué au Québec et appelle vigoureusement à la tenue d'un débat public sur cette question. Fischler lève le voile sur un assemblage confus de corps publics et d'institutions qui sont impliqués—ou, à tout le moins, qui estiment *devoir* être impliqués—dans la planification de projets routiers d'envergure. La problématique de la re-construction de l'échangeur Turcot, selon Fischler, relève tout à la fois d'un problème technique, d'un problème urbain et d'un problème politique.

**Pierre Gauthier** traite dans le chapitre 6 de la problématique de l'intégration spatiale des infrastructures autoroutières urbaines dans les quartiers densément peuplés. Il présente ensuite une méthode utile à l'évaluation des impacts d'une autoroute sur la forme urbaine et la qualité de vie des populations riveraines. Il avance que dans l'hypothèse où la construction d'une autoroute urbaine ne saurait être évitée, une telle méthode peut à tout le moins servir à l'établissement de critères destinés à guider nos actions.

Une proposition de requalification du secteur Cabot à Côte-Saint-Paul est présentée en guise d'illustration des avantages de l'approche proposée. **Elham Ghamoushi, Jonathan Moorman, Erika Brown** et **Munaf Von Rudloff** examinent d'une manière innovatrice au chapitre 7 comment la proposition du MTQ et celle du duo Brisset et Moorman (présentée au chapitre 2) se conforment aux politiques actuelles en matière environnementale et de transport. Les auteurs analysent treize énoncés de politique censés guider les actions gouvernementales dans la région montréalaise en vue d'y atteindre un niveau accru de « soutenabilité ». Ils relèvent 124 objectifs relatifs au bruit, aux transports et aux aspects socio-économiques en rapport avec la reconstruction de l'échangeur Turcot, pour évaluer ensuite les impacts respectifs des deux propositions à l'égard de ces critères et des directives politiques. Les auteurs concluent que l'approche promouvant le développement du transport en commun (Brisset·et Moorman) est beaucoup plus en accord avec les objectifs des politiques de développement soutenable que le projet du MTQ.

Il existe un impressionnant corpus recensant les effets négatifs sur le bien-être physique et psychologique des personnes qui habitent à proximité des autoroutes. Dans le chapitre 8, **Meaghan Ferguson, Robert Moriarity, Frédéric Gagnon** et **Melanie McCavour** présentent un bref ·survol de la recherche scientifique récente quant aux effets du transport routier sur la santé, en insistant sur les impacts de la pollution de l'air et du bruit. Les impacts les plus prononcés s'observent à une distance de moins de 200 mètres de l'autoroute. Les auteurs concluent que l'évaluation des risques pour la santé des secteurs limitrophes du carrefour autoroutier et des autoroutes 720 et 15, nécessite une recherche plus approfondie que ce dont nous disposons à ce jour.

Dans le chapitre 9, **Jacob Larsen** examine de près l'argument à l'effet que les autoroutes dans les quartiers centraux sont nécessaires pour soutenir l'économie urbaine, notamment à l'égard des besoins de transport par camion. Larsen passe en revue la manière dont Montréal a traité historiquement la question du transport sur route des marchandises et, sur la base des conclusions de la Commission Nicolet de 2003, nous fait valoir la possibilité de détourner bon nombre de camions des corridors autoroutiers localisés dans les zones densément peuplées du Sud-Ouest de Montréal. Cet article nous aide à comprendre le genre de mesures incitatives et coercitives qui peuvent être employées pour encourager un changement de comportement de la part des camionneurs.

**Ian Lockwood** et **Joel Mann** partagent leurs expériences et leurs réflexions sur les autoroutes, le transport et l'urbanisme dans le chapitre 10.

Ils procèdent à un survol de l'origine et des impacts des stratégies actuelles en matière de transport, et traitent des enjeux qui pourraient ou devraient affecter ce domaine dans le futur.

★ ★ ★

Dans un souci de documentation historique, nous reproduisons quelques documents-clés qui ont alimenté les débats sur l'échangeur Turcot et les projets du MTQ au cours des derniers mois.

Cet ouvrage se conclut sur un bref chapitre de la plume des co-directeurs.

★ ★ ★

Les co-directeurs désirent exprimer leurs remerciements à Lisa Bornstein, Jill Prescesky et Grace Keenan Prince pour leur aide à l'édition et à la révision des chapitres de ce livre, ainsi qu'à Zack Taylor pour la mise en page et l'aide à la révision finale. Cet ouvrage s'inscrit dans la foulée du travail effectué dans le cadre du projet "Mégaprojets au service de communautés" financé par le Conseil de recherche en sciences humaines du Canada à travers le programme Alliance de recherche universités-communautés (ARUC).

★ ★ ★

Au moment où cet ouvrage s'apprête à être mis sous presse, le Conseil régional de l'environnement de Montréal nous annonce que le BAPE a acquiescé à sa demande et que ce dernier tiendra une séance spéciale portant sur les projets de transports en commun présentement en préparation dans le corridor Est–Ouest de Montréal. Ces projets incluent la très attendue navette ferroviaire de l'aéroport, le projet de « tram-train » de Lachine et la création de voies d'autobus réservées. Le chapitre 2 montre comment lesdits projets, s'ils sont bien menés et mis en œuvre conjointement, pourront réduire le volume de circulation de manière significative dans le corridor Est–Ouest. D'autres chapitres de ce livre contiennent des informations et des perspectives qui sont de nature à nourrir le débat public à l'occasion des audiences du BAPE ou dans d'autres forums. Nous souhaitons en outre que le matériel inclus dans notre *Montréal à la croisée des chemins/Montréal at the Crossroads*, quoi que focalisant sur Montréal et sur l'échangeur Turcot, enrichira les discussions qui ont lieu en d'autres villes, où des infrastructures autoroutières parvenues au terme de leur durée de vie utile doivent être reconstruites, rénovées, re-développées ou dont l'existence même est questionnée. Alors que les modèles, aujourd'hui datés, qui ont présidé à l'organisation des transports, du développement urbain et dans une certaine mesure, de la vie urbaine elle-même, font l'objet de remises en cause, de tels débats sont essentiels à l'établissement d'une vision et de visées communes pour l'avenir de nos villes.

*Pierre Gauthier*
*Jochen Jaeger*
*Jason Prince*

Montréal, mai 2009

# Montreal at the Crossroads

As this book hits the presses, Montreal finds itself engulfed in a debate about the Turcot Interchange.

One of the largest interchanges of its kind in Quebec, this elevated highway intersection handles over 280 000 vehicles a day, moving cars, trucks and buses to their destinations in different quadrants of the city. The Turcot was built in the 1960s, just in time for Montreal's World Exposition. Now over 40 years old, the structure is showing its age. What is to be done?

Engineers at Quebec's Ministry of Transportation (MTQ) have been hard at work for years coming up with a proposal for replacing this crumbling infrastructure. Their solution is to replace this elevated structure with a new structure, built on embankments.

It is important to understand that the MTQ proposal includes not just the messy knot of the Turcot Interchange itself, tying Autoroute 15 (North-South) to the 720 (East-West), but two other intersections to the West and South, as well as several kilometres of connecting highway, some elevated and some already on the ground (see colour images in the centre of this book).

Increasing the road capacity slightly, the MTQ expects traffic volumes in the entire network (the Turcot proper and all its connected parts) to increase from 332,000 vehicles per day to 347,000 vehicles per day by 2016. They do not anticipate ambient pollution to increase, because (they argue) new cars will be more and more efficient and pollute less. So, even with more cars, there will be less pollution.

There appears to be a consensus among critics on one important thing: whatever happens to the Turcot, the North-South structures and traffic volumes should probably be maintained as there are no alternatives for the foreseeable future to replace this critical segment of the national highway network. However, the relevance of rebuilding the East-West route along the 720, in the form of a highway, is much more debatable. It principally accommodates commuters, who could be moved using mass-transit. Indeed, they are very "clientele" recognised as potential mass-transit users in municipal initiatives to strengthen transit systems and dismantle intrusive and environmentally unsound urban highways.

The new structure, it is argued, will be easier and cheaper to maintain, and hence will relieve future generations from a financial burden (a key component of the MTQ's definition of "sustainable development"). The new intersection will be built to modern highway standards, with shoulders, and hence safer.

Critics argue that such a rationale misses the mark at numerous levels. The climate-change crisis and the health issues associated with our dependence on the car require more vision, and more immediate action. First, wherever possible, more sustainable modes of transportation need to be offered as an alternative to the automobile. Second, densely built inner-city neighbourhoods that were decimated by the construction of urban highways in the 1960s and 70s are the very areas into which urban development should now be funnelled. This is particularly true of the former inner-city sites of production that now sit empty. The current MTQ proposal does nothing to fix the errors of the past by investing in improving the quality of life in Montreal, and would sacrifice, for generations to come, the development potential of a critical part of the city's centre.

Moreover, critics argue that embankments are a rural solution to an urban problem. Embankments have no place in the city. The MTQ proposal results in walls that divide neighbourhoods, create dead-end streets and small isolated pockets of near-useless land, with sterilized land values and diminished tax revenue for the city, and entail a highway with a tripled footprint (an embankment on either side of it), wasting additional hectares of valuable inner city land.

In the MTQ proposal, the new road must be built beside the existing one in order to maintain traffic flow during construction. Once completed, traffic can be rerouted and the old structure demolished. The additional space required during construction means the expropriation and demolition of nearly 200 units of good quality housing in a tiny sub-neighbourhood of Saint Henri, called the *Village des tanneries*. (Anyone taking the ramp going East off the Décarie Expressway can get a glimpse into the curtained bedroom of a loft apartment, a mere 30 feet from the curving concrete. Some of the housing is that close!)

The Minister has offered to build affordable replacement units "nearby" and to financially compensate tenants and homeowners, as stipulated in Quebec law. But many residents may not be eligible for these new units (the programs are very restrictive). Residents are understandably upset at the partial demolition of their neighbourhood, and what is left of it may not survive the shock.

The MTQ promises to plant dense vegetation on the embankments: residents are sceptical, as a survey of Quebec's highways shows nothing comparable to the beautiful renderings provided by the MTQ during public consultations.

The cost: the MTQ estimated in 2006 that their embankment project for the Turcot will cost 1.5 billion dollars. No one takes this estimate seriously, for reasons only a Montrealer can really understand. One recent example of escalating costs will suffice to make this point. Cost estimates for the Notre Dame highway project, just east of downtown Montreal, jumped from $750,000 in 2006, when the project was approved, to $1.5 billion, in less than two years. The MTQ acted swiftly to halt the project and it is currently under review again. The Minister claims this will not happen with the Turcot projections—they have budgeted for "les imprévus", the unexpected—but many remain sceptical.

★   ★   ★

Hundreds of thousands of people drive through the Turcot Interchange every day. Hundreds of thousands of others hear about it on morning and afternoon radio. For some, those who don't drive, it is a mysterious place, living only in their imaginations, a place with stalled cars in the left lane, or perhaps "easy going". For others, leaving each morning in their car for work, it is an almost necessary route. And a gamble. Will I get stuck today? These Montrealers listen like hawks to the radio, trying to guess the best way to navigate the city to their destination.

★   ★   ★

Something must be done about the Turcot, all will agree.

The MTQ claims to have reviewed the renovation option in 2004, and rejected it as too costly, requiring more time than the demolition and reconstruction option they ultimately adopted. But the documents that show this analysis have not been submitted for public consultation and remain internal.

Since the collapse of a Montreal overpass, de la Concorde, in 2006, which killed five people and injured many others, Quebec has been extra vigilant. An article in the French daily paper *Le Devoir* dated October 2008 calls the Turcot Interchange "la structure la plus surveillée au Québec." Literally hundreds of repairs have been made on the structure in the past 5 years. Drivers can't avoid glancing up nervously at the steel mesh tacked onto the undercarriage of the concrete ribbons as they drive under them.

Many drivers sense its imminent collapse. But the 24 hour inspections, costing millions each year, help inspire confidence. People still use it. Can the Turcot be repaired? Should we reduce its capacity and move people differently, on trains, trams and buses? Should we simply dismantle it?

Should it be preserved as a monument to Modernity?

★ ★ ★

This volume is a contribution to the debate.

It is vitally important to conduct this debate now. The MTQ hopes to begin construction on their project before the end of 2009. Public hearings on the project have begun in earnest (accelerated by a government eager to get the project moving), managed by the *Bureau d'audiences publiques sur l'environnement* (BAPE). Residents, groups and even the experts are overwhelmed by the volume of information contained in the Environmental Impact Assessment, technically the object of public hearings, while at the same time arguing that they do not have enough information (or more precisely the right *kind* of information) to make an informed opinion.

The *Conseil régional de l'environnement de Montréal* (CREM), an environmental lobby group with considerable influence, has argued—and to some extent, successfully—for a broader debate than that allowed by the BAPE rules. A letter that the CREM submitted to Line Beauchamp, minister responsible for Sustainable Development, the Environment and Parks, is included as Appendix B. Another group has launched a legal appeal, arguing that hearings must be delayed until we have additional information. The Table de travail Turcot, a group of urban planners, academics and community groups, have argued that the MTQ is a technical solution to the wrong question (how do we move this volume of cars at this speed to these places?) and what is needed is a much more general debate—leading to a more holistic solution— about how to achieve mobility goals inside a better-designed city. We include their appeal in Appendix A.

Many cities and their residents the world over are questioning how best to deal with aging infrastructure built for the automobile, during the cheap fuel era.

Montreal is no exception, and must not shirk from this debate.

This volume brings together a range of viewpoints that we hope enriches the discussion and suggests new ways of thinking about our city and its future.

We hope that this modest contribution stimulates debate and brings about the best possible Montreal, for all residents.

★ ★ ★

This book is structured around the major debates and issues that have been raised in the past year about the Turcot, in dozens of networks across the city. Some of the material has been written by seasoned experts and academics, others by students who have conducted careful research on focused areas, and yet others by emerging community leaders.

This book, then, contains a multiplicity of voices. Let us guide you through these chapters and introduce you to the concerns.

Montreal's current transport system is characterized by both automobile dependence and transit orientation. In Chapter 1, **Jeff Kenworthy** and **Craig Townsend** argue that rebuilding the Turcot Interchange for cars, while neglecting mass transit, will increase automobile dependence and weaken transit competitiveness. The authors provide data comparing Montreal's transportation system to other cities around the world and conclude that Montreal needs improved transit and does not need more high-capacity roads.

Alternatives to the current Ministère des Transports du Québec (MTQ) proposal are offered. In Chapter 2, **Pierre Brisset** and **Jonathan Moorman** advocate a transit-oriented vision for the Turcot interchange consistent with the Montreal Transport Plan of 2008. Their proposal combines transit improvements and disincentives for car users to reduce the number of drivers commuting from the West Island and inner city neighbourhoods to the center of Montreal. It will then be possible to redesign the interchange and reduce the size of the Ville-Marie Highway in three phases. The authors propose an exciting long-term vision where the Ville-Marie is removed from the Turcot to Atwater and the Turcot Interchange is converted into a three-way junction.

**Pieter Sijpkes** contributes a vivid and impassioned chapter to the book. Sijkpes, in Chapter 3, argues that the Minister's proposal for the Turcot is mistaken on two counts. First, rather than scrapping the concrete structure, we should focus on what still works. The roadbed could be supported with a steel understructure. This approach would prolong the life of the structure for years to come, cost less, enable ongoing inspection and repairs, and—crucially neither disrupt traffic flow during construction, nor require demolition of homes. Second, he argues against the MTQ proposal to relocate highway 20 and the CN rail lines up against the Falaise Saint Jacques. Instead, Sijpkes proposes to broaden this linear park, restore the St. Pierre River and screen the park from highway 20 with careful landscaping.

**Patrick Asch** develops this park concept further in Chapter 4, arguing that the Falaise Saint Jacques is an undiscovered jewel among Montreal's green spaces. It rivals Mount Royal Park in size, and with the addition of several strategic green links along existing rights of way, this linear park could become the backbone connecting over 500 hectares of green spaces in the South-West of the island. Asch also makes a bold proposal to swap contested land in nearby Meadowbrook for an equivalent number of hectares in the Turcot Yards, to enable residential redevelopment. Asch warns us that the MTQ proposal to move the highway and railways against the Falaise will destroy this green space, eliminating its natural, social and even economic function in our City.

In Chapter 5, **Raphael Fischler** reminds us how complicated urban planning is in Quebec and urges public debate on the issue. Fischler sheds light on the complex and overlapping jurisdictions and institutions that are involved—or feel they *should* be involved— in planning major road projects. The "problem" of the Turcot Interchange, for Fischler, is at once a technical problem, an urban problem, and a political problem.

**Pierre Gauthier** discusses the problems associated with the spatial integration of urban highways in densely populated urban areas in Chapter 6. He introduces a method that could help to evaluate the impacts of highways on the urban form and the quality of life of the neighbouring populations. He argues that if a highway construction cannot be avoided, use of such a method could at least produce criteria to guide our actions. A redevelopment proposal for the Cabot area of Côte-Saint-Paul is used to exemplify the relevance of the approach.

**Elham Ghamoushi, Jonathan Moorman, Erika Brown** and **Munaf Von Rudloff**, in a thought-provoking chapter (Chapter 7), examine how well the MTQ plan for the Turcot and the Brisset-Moorman alternative (Chapter 2) conform with current environmental and transport policies. The authors analyze 13 policy documents intended to guide decision-making in the Montreal region towards a higher level of sustainability. They identify 124 goals related to noise, transport and socioeconomic aspects of the Turcot's reconstruction. They then assess the impacts of the two alternative proposals against these policy directives. The authors conclude that the public-transit approach is far more consistent with sustainable policy goals than the current MTQ proposal.

There is an impressive body of evidence that traffic has deleterious effects on the physical and mental well-being of people who live near highways. In Chapter 8, **Meaghan Ferguson, Robert Moriarity, Frederic Gagnon** and **Melanie McCavour** provide a brief overview of the current scientific research on health effects of road traffic with an emphasis on air pollution and noise. Impacts are particularly pronounced within a distance of 200 metres from highways. They conclude that assessment of the health risks requires more detailed research and measurement near the Turcot Interchange and adjacent highways 720 and 15 than conducted to date.

In Chapter 9, **Jacob Larsen** tackles the argument that inner city highways—and specifically the trucking they enable—are needed to support the urban economy. Larsen reviews how Montreal has historically approached the transport of goods and, highlighting the findings from the Nicolet Commission (2003), shows us how one could reroute many trucks away from the densely populated corridors in the south-west of the city. This article helps us to understand the kinds of incentives and disincentives we can employ to encourage change in driving behaviour.

In Chapter 10, **Ian Lockwood** and **Joel Mann** share their experience and thinking about highways, transportation and urban planning. They provide an overview of how today's transportation strategy came to be, how it has delivered, and the major issues that could and should cause it to change in the future.

For historical interest, in the last part of the book we include key political documents that have galvanized debate in the past months on the Turcot Interchange and the MTQ plans, as described above.

The book concludes with an afterword from the editors.

★ ★ ★

Special thanks to Lisa Bornstein, Jill Prescesky and Grace Keenan Prince for commenting on these chapters, and also to Zack Taylor for assistance with the layout and final editing. This volume contributes to the work undertaken by the community-university research alliance (CURA): *Making Mega-Projects Work for Communities*, funded by the Social Sciences and Humanities Research Council (SSHRC).

★ ★ ★

As this book goes to press, the Conseil régional de l'environnement de Montréal is announcing that the BAPE will indeed hold special hearings on the major transport projects on the drawing table for the East-West corridor in Montreal. These projects include the long-awaited airport train shuttle; the Lachine tram-train; and reserved bus lanes. Other chapters of this book include information and perspectives that inform public debate, whether in the BAPE hearings or in other forums. In addition, it is our hope that the material contained in *Montreal at the Crossroads*, though focused on Montreal and the Turcot Interchange, will enrich similar discussions in other cities where aging infrastructure must be rebuilt, refurbished, redesigned or rethought. As our older models of transport, urban development and urban life, are called into question, such debate is an essential part of establishing a common vision, direction and set of actions for our future cities.

*Pierre Gauthier*
*Jochen Jaeger*
*Jason Prince*

Montreal, May 2009

# CHAPTER ONE

# Montreal's Dualistic Transport Character: Why Montreal Needs Upgraded Transit and Not More High Capacity Roads

*Jeff Kenworthy and Craig Townsend*

The Turcot represents a crucial decision-making moment for Montreal. Montreal exhibits a dualistic transport character of automobile dependence and transit orientation. This character resulted from one century of urban development around streetcars, followed by almost half a century of urban development around freeways and arterial roads, together with pockets of development around an underground railway. Plans to rebuild the Turcot Interchange, a key segment of Montreal's freeway infrastructure, in the absence of commensurate investments in rapid transit will increase automobile dependence and imperil transit competitiveness. Data illustrating Montreal's transportation character and recent changes are presented with comparisons to other cities around the world.

Montreal has traditionally been, and remains, Canada's most transit-oriented metropolis. This transit orientation was established over the course of a century, between 1861 and 1959, when electric streetcars provided transportation along densely built-up, mixed-use streets (Angus, 1971). During one quarter of the streetcar era, between 1906 and 1933, much of Montreal's "European-style" central and inner suburban areas was built. Since the 1950s most of metropolitan Montreal's new growth has sprawled along with massive freeways built in the 1950s and 1960s; as a result, this new growth has been automobile dependent. Countering Montreal's postwar automobile orientation was the decision of Mayor Jean Drapeau to build underground rail rapid transit, the Metro, in the 1960s. The Metro (now 65km and 68 stations) provided fast, comfortable, and frequent rapid transit service, which together with proactive and sympathetic land use planning led to transit-oriented development in the central city and some other key locations. Montreal's constrained growth in automobile dependence and persistent strength in transit orientation during an age in which most cities became more automobile-oriented were directly linked to these infrastructure investments.

At the beginning of the twenty-first century Montreal exhibits a dualistic character, both automobile-oriented and transit-oriented, that distinguishes it from other metropolitan areas. The Turcot Interchange opened in the same year as the Metro. The provincial transportation ministry's proposal to rebuild the interchange, without commensurate public transit improvements, combined with ongoing freeway-building across the metropolitan area will push Montreal further toward the automobile dependent growth that was initiated in the 1950s. In this sense, Montreal's post-war story is similar to that which has been repeated again and again around the world, and which has only been significantly broken in places that have removed freeways and parking (e.g. Portland, Oregon and Copenhagen, Denmark).

The potential impacts of the proposed Turcot Interchange reconstruction are considered through the presentation of comparative panel data from 1996 and 2006. Montreal is analysed in relation to a selection of metropolitan areas in Canada, the USA, Australia and Europe. The data are drawn from a longstanding and internationally acknowledged research effort comparing the level of automobile dependence in cities (e.g. Kenworthy and Laube 1999, 2001). Data for 2006 are from a yet-to-be published update of this work.[1]

## Transport Infrastructure

### Freeways

Metropolitan Montreal has abundant freeways, even when compared to US metropolitan areas. In 2006 Montreal had 0.156 metres of freeway per person, an increase of 8% since 1996. Of a sample of 25 metropolitan areas worldwide, only six had more per capita freeway provision. One was Calgary and the other five cities were in the USA (Denver, Houston, Phoenix, San Diego and Washington). Some metropolitan areas renowned for their freeway orientation actually had less freeway length per capita than Montreal; for example, Los Angeles had only 0.090 metres per person and Atlanta had only 0.152 metres per person (i.e. Montreal had 73% and 3% higher freeway provision respectively). However, Montreal had less wealth than these regions, raising the question of how residents of Montreal and its suburbs will afford more and more freeway per person. In 2006, the per capita GDP of the Montreal metropolitan area was $26,685 compared to an average 2005 GDP for the ten US metropolitan areas in this comparison of $42,547 (both figures are 1995 US dollar equivalents).

Reserved rights-of-way for transit

Railways and bus-only lanes provide the infrastructure necessary to enable services which can compete with cars in locations served by freeways. In terms of transit infrastructure with reserved rights-of-way for transit, Montreal performed well. In 2006 Montreal had 119.2 metres of reserved transit right-of-way per 1000 people, up almost 100% between 1996 and 2006. In comparison with 21 other metropolitan areas, there were 15 that had less per capita reserved right-of-way than Montreal, averaging only 66.5 metres per person and having risen by 25.9% over the 10 years (Atlanta, Calgary, Denver, Houston, Los Angeles, Melbourne, New York, Ottawa, Perth, Phoenix, San Diego, San Francisco, Toronto, Vancouver, Washington DC). For the 6 metropolitan areas where reserved rights-of-way were higher than in Montreal, these averaged 212.6 metres per person or some 78% higher than in Montreal (Berlin, Brisbane, Chicago, Hamburg, Munich, Sydney). Amongst these metropolitan areas the increase in this factor was on average 12.4% since the mid-1990s. According to this statistic, Montreal had a healthy level of public transport protected rights-of-way and this rose proportionately more than other metropolitan areas over the decade under consideration.

This 1996-2006 increase in rapid transit infrastructure was the result of the creation of the *Agence métropolitaine de transport* (AMT) suburban train lines, which began operations in 1996. While these trains are highly popular among residents of automobile-oriented suburbs, the impact of the infrastructure in attenuating the increase in automobile dependence has been minimal. Some of this has to do with the use of land around the stations. Most of the AMT suburban train stations, spaced far apart and limited in number, are surrounded by surface parking, provided at a cost of approximately $5,000 per stall. Concerns about the impact of surface parking and the commuter trains on suburban sprawl led in the early 2000s to a program to consider options for Transit-Oriented Development (TOD), and to assess whether the suburban trains were actually encouraging automobile dependent sprawl (Rivard and Le Colletter, 2006).

In 2006 Montreal still had more freeway than rapid transit line (156 metres of freeway per 1000 people compared to 119 metres per 1000 people of reserved transit right-of-way), so that the orientation in terms of high speed transport infrastructure was still with the automobile. The 2007 opening of the $0.8 billion, 7.5km Laval Metro extension added only around 2 metres per 1000 people to the public transport right-of-way and thus the ratio would hardly have changed.

## Car-Orientation

Car ownership

In comparison with other North American metropolitan areas, Montreal in 2006 had a very low level of car ownership: 446 cars per 1000 people. All US metropolitan areas far exceeded this figure, except for the most transit-oriented metropolitan area in the USA, New York, which coincidentally had an identical level of car ownership to Montreal. The other nine US metropolitan areas in this study averaged in 2005 an extraordinary 660 cars per 1000 people or 48% more than Montreal. Calgary had 632 cars per 1000 people, Vancouver 506, and Toronto 485, all considerably higher than Montreal. Even some European cities exceeded the Montreal region in car ownership: Frankfurt had 512 cars per 1000 people, Hamburg 484, Munich 531 and Zurich 515.

Car use

The critical item describing the car orientation of a city is not so much car ownership as car use. In this respect Montreal distinguishes itself as being a North American city with relatively modest levels of car use. In 2006 car use in the Montreal region was 5,333 car kilometres per capita, a small decline of 1.8% from 1996. To put this in perspective, in 2006 the four Australian metropolitan areas averaged 8,700 km per capita or 63% higher than Montreal and representing an increase over 1996 of 14%. In Calgary car use was 8,362 km per capita, a 1.5% increase from 1996 and in Vancouver 6,971 km, a 3.3% increase from 1996, and both obviously much higher than Montreal. Toronto on the other hand, another of Canada's more transit-oriented metropolitan areas, had car use similar to Montreal at 5,020 km per capita, but representing an 8.6% decline from 1996. Eight US cities (Atlanta, Chicago, Denver, Houston, Los Angeles, Phoenix, San Diego and San Francisco) averaged 13,471 km per capita or 2.5 times more car use than Montreal. This figure represents an average decline over the ten years of 0.8%. Of the two European cities currently available, car use per capita was as follows: Zurich: 5,119 km, or 4% below Montreal and Frankfurt 5,991 km or 12% more than Montreal. Frankfurt's car use over the ten years rose 14.2%, while Zurich declined 4.7%.

**Transit service and use**

Transit service level

While Montreal had a relatively high level of transit infrastructure, transit service levels were relatively low. Montreal had 2,048 annual seat kilometres of service per capita offered to the people of the region in 2006, but this was 14% less than in 1996. This reflects the achievement of an increase in rapid transit through suburban train services which are infrequent and low capacity (the AMT trains carry approximately 64,000 riders per weekday, compared with approximately 845,000 on the Metro) in comparison with inner city buses and Metro lines. In contrast, fourteen other metropolitan areas had a higher level of service provision, averaging 3,705 seat km per capita in 2006, or a very large 81% higher provision of transit service than Montreal. For these cities, transit service per capita expanded 12% over the decade, whereas Montreal declined 14%.

Metropolitan areas with lower transit provision than Montreal in 2006, averaging 1,050 seat kilometres per capita, or about half that of Montreal, were all highly auto-oriented US cities (Atlanta, Denver, Houston, Los Angeles, Phoenix and San Diego). Nevertheless, even these cities had expanded their transit service provision over the decade by 13%. Not only is Montreal's transit service provision relatively poor compared to other metropolitan areas, except for that in the most automobile-oriented US metropolitan areas, it has also declined while others have increased their level of transit provision. The Province of Quebec's proposed 51 km, $0.4 billion line to Repentigny would provide service to about 5,000 weekday travellers, and would further diminish Montreal's total transit service levels relative to infrastructure length.

Transit use

In terms of transit use, measured by annual boardings per capita, Montreal fared quite well. In 2006, Montreal had 206 boardings per capita, which is the highest use of transit per person of any North American metropolitan area. Notwithstanding this achievement, it is exactly the same as it was in 1996, indicating that in spite of a doubling of rapid transit infrastructure through suburban railways, it has not been able to gain ground over the automobile. The only two cities outside of Europe that came close to Montreal in transit use were New York at 168 trips per capita and Toronto at 154 trips per capita.

By way of contrast, four Australian metropolitan areas averaged 96 boardings per capita, but up by 6% from 1996. The other four Canadian cities in the sample averaged 137 boardings per capita, up by 10% from 124 boardings per capita in 1996, while the US cities averaged 67 boardings per capita, up 11%. The only metropolitan areas that exceeded Montreal were in Europe, averaging 418 trips per capita or over double that of Montreal. And these European cities increased their per capita transit ridership by 23% over the decade, whereas Montreal was stagnant.

### Estimated Impacts of the Turcot Interchange Reconstruction Project

The data presented suggest that in some ways Montreal's transport system is public transport-oriented, like those in Europe, while in other ways it is like an automobile dependent North American city. Montreal's transport system has a dualistic or oscillating character. The automobile-dominated character of Montreal has been made possible by large transport infrastructure projects such as the Turcot Interchange. While re-building the Turcot Interchange will not make a difference to the length of freeway per capita in metropolitan Montreal, the $1.5 billion budget will reduce available funding for other transport projects and will therefore reinforce Montreal's automobile oriented character. This will be pronounced if there is a lack of similarly scaled infrastructure improvements to rapid transit infrastructure with high service levels. Building of the Turcot Interchange in the late 1960s coincided with the building of the Metro. Where is the commensurate investment in new mass-transit infrastructure?

When the low service levels and automobile oriented land use impacts of Montreal's suburban trains are considered, it is clear that Montreal is quickly falling behind most peer metropolitan areas in terms of the provision of rapid transit service. Ironically, Montreal's transit stagnation and impending decline is occurring just as other North American metropolitan areas, generally lacking the favourable built form conditions found in Montreal, are seeking to shift people to transit through massive investments in rapid transit infrastructure that will offer high service levels. In 2009 the Province of Ontario committed to over $7 billion in transit infrastructure improvements while the Province of British Columbia committed to building a $1.4 billion light rail line, in the same year that the 20km, $1.5 billion Canada Line to Vancouver International Airport and Richmond will open for service.

In summary, Montreal is at a crossroads: governments must decide whether to continue to invest in high capacity road infrastructure at the expense of significant transit expansion, or to invest more money in developing transit, in which Montreal already has a comparative but slipping advantage. It is difficult to see how, under these conditions and given Montreal's largely stagnant public transit performance, Quebec's goal (Quebec, 2006) to increase public transit use by 12% by 2012 will be reached. Since Montreal already ranks very highly in its freeway infrastructure provision while transit infrastructure seriously lags behind it, and transit service levels are very poor, the choice would seem to be fairly obvious if Montreal's sustainability credentials and liveability for its citizens are to be strengthened.

## Notes

1. The authors wish to sincerely acknowledge the generous support of the Helen and William Mazer Foundation of New Jersey without whose research funds to support the update of the global cities database, this chapter would not have been possible. The authors also wish to gratefully acknowledge the research assistance of Ms Monika Brunetti in assembling the 2006 data in this chapter.

## References

Angus, F.F. (1971). *Remember Montreal's Streetcars!* Montreal: Canadian Railroad Historical Association. 1-8.

Kenworthy, J.R. and Laube, F.B. (1999) *An International Sourcebook of Automobile Dependence in Cities, 1960-1990.* University Press of Colorado, Niwot, Boulder, Colorado. 704 pp.

Kenworthy, J. and Laube, F. (2001) *The Millennium Cities Database for Sustainable Transport.* (CDROM Database) International Association of Public Transport, (UITP), Brussels and Institute for Sustainability and Technology Policy (ISTP), Perth.

Quebec. (2006). Better Choices for Citizens : Quebec Policy Respecting Public Transit.

Rivard, G. et Le Colletter, E. (2006) « Relations entre le développement récent du réseau de trains de banlieue et l'étalement urbain dans la région métropolitaine de Montréal ». AMT. octobre.

# CHAPTER TWO

# A Transit-Oriented Vision for the Turcot Interchange: Making Highway Reconstruction Compatible with Sustainability

*Pierre Brisset and Jonathan Moorman*

The Turcot Interchange was built to sustain car-oriented urban development. In urgent need of reconstruction, the Turcot Interchange should be redesigned in light of sustainable transit-oriented goals, such as those enunciated in the Montreal Transport Plan 2008. Implementation of major transit projects and adoption of car reduction policies would make possible a reduction in the interchange's traffic capacity and physical size, and diminish its impact on adjacent neighbourhoods. Land use planning, transit planning and parking or tolling policies must be integrated to the infrastructure plan in order to help reduce car dependency and improve health and quality of life for Montrealers. The proposal presented here combines transit improvements and disincentives for car users in order to reduce the number of drivers commuting from the West Island to the center of Montreal. By reducing this number it is possible to redesign the interchange to sustain this new transportation approach. Most importantly, we propose to reduce the capacity of the Ville Marie Expressway in three phases, converting the Turcot Interchange into a three-way junction.

## Introduction

Rather than simply rebuilding a structure designed to meet a 1950's vision of the city, the Turcot project could be the first step in a lasting transformation of how people commute and travel in Montreal. It could also be the beginning of a new form of urban development, a catalyst for the creation of green transit-oriented neighbourhoods.

The urgent need for reconstruction of the Turcot Interchange and the project undertaken by the Quebec Ministry of Transport (MTQ) represent a unique opportunity for the citizens and leaders of Montreal. Many stakeholders have diverse interests in the future of this infrastructure mega-project; the challenge for the governments of Montreal and the province of Quebec will be to choose a course of action which benefits the city and the province in a long term, sustainable way.

The Turcot Interchange plays a vital role in the Greater Montreal transportation system, but also supports a high level of automobile dependency. It must continue to provide efficient, safe, and reliable forms for transporting people and goods. It must also conform to the long term, sustainability-oriented vision adopted by the City of Montreal both in its Master Plan and Transport Plan, and to the environmental vision the MTQ has set out in its policy.

## Planning for Sustainability

Municipalities and government agencies are focusing more and more on sustainable development and on the integration of economic, environmental and social equity goals into the planning process (Carmona & Sieh, 2008; Handy, 2006b). In Quebec, both the provincial and the Montreal municipal governments have adopted policies aiming towards sustainable development (see chapter 8 by Ghamoushi et al.).

Economic growth over the last two decades has produced a growing need for the efficient transportation of people and goods. Automobile use has steadily increased and suburbanization has followed. Between 1987 and 1998 the number of auto trips increased by 30% in the Montreal region, while transit's modal share dropped from 37% in 1982 to 21% in 2003 (AMT, 2003a) (see chapter 4 by Kenworthy and Townsend).

This trend has very negative consequences on health and the environment. Global warming from car emissions is among the most serious of these. In 1991, 27% of all energy consumed in Quebec was transport-related; of this, 83% was consumed by automobiles (MTQ, 1994). Transport contributes to 47% of Montreal's greenhouse gas emissions, 85% of NOx emissions, and 30% of all airborne particulates (Drouin, 2008; Drouin, et al., 2006; DSP, 2006).

In addition, air pollution from automobile use is a serious health issue: air pollution caused 1500 premature deaths in Montreal in 2002 (Drouin, 2008; DSP, 2006). Cardio-respiratory disease, road injuries, obesity, diabetes, and exclusion of persons with reduced mobility are all identified as direct results of the changes in air quality, climate, road safety and accessibility which follow increased automobile use (Drouin, 2008; Drouin, et al., 2006; DSP, 2006; see also chapter 7 by Ferguson et al.).

Transportation policies adopted by both the MTQ and the City of Montreal assert the importance of reducing car dependency and place transit at the heart of sustainable development strategies. These policies also highlight the interaction between the land use and transportation systems as well as the importance of integrating land use planning with transportation infrastructure projects.

## Transit and Sustainability

The MTQ adopted its environmental policy in September 1992. It makes sustainable development its guiding principle and places environmental concerns at the heart of decision-making alongside economic issues. Within its environmental policy document, the MTQ recognizes the fundamental link between land use planning and transportation. The document states that urban sprawl and less dense suburban areas favour car usage; this leads to increased noise and air pollution in central city areas. A transportation plan which integrates land-use planning could reduce pollution through the harmonious growth of urban development and transit. The MTQ also asserts that diminishing the negative effects of car use can only effectively be achieved by replacing cars with transit (MTQ, 1994).

The MTQ's environmental policy represents a financial and strategic commitment to the implementation and improvement of public transit. Developing public transit is a priority in order to meet today's needs without compromising those of future generations (MTQ, 2006).

Although transit ridership has slowly been declining since the 1950's, that trend is now reversing (AMT, 2003b),and transit use in Montreal has stabilized since 1998 (City of Montreal, 2003). While transit ridership declined by about 1% a year between 1986 and 1995, it increased 1.7% annually between 1996 and 2002 (AMT, 2003b). In fact, the annual growth in car trips (+0.9 %) was lower than the annual growth in transit trips (+1.6 %) for the year 2003 (see chapter 4 by Kenworthy and Townsend).

Economic and demographic growth on the island of Montreal as well as changes in service levels of transit to better meet needs may explain these results (City of Montreal, 2003). In order to meet Kyoto objectives, transit ridership should increase by 5% yearly (AMT, 2003a). Aggressive initiatives are needed to prioritise public transit so it can adequately meet transportation goals and be competitive with automobile use.

## The Transit Metropolis

Following the merger of the municipalities on the island of Montreal in 2002, Mayor Gérald Tremblay held the Montreal Summit to determine development priorities. The Montreal Master Plan and the Montreal Transport Plan 2008 were the result. Both these plans are based on sustainable development principles and have the goal of improving Montrealers' quality of life. They are future-oriented and environmentally focused; in fact, one of the stated aims of the Master Plan is "to guarantee that future development in Montreal will rest firmly on the principles of sustainable development". Explicitly-stated goals are to "maintain the quality of established living environments" and to "ensure the positive contribution of large transportation infrastructure to the urban landscape" (City of Montreal, 2004). The Master Plan addresses issues such as enhancing transportation efficiency and effectiveness, protecting natural areas, and improving environmental conditions.

The focus of the Montreal Transport Plan 2008 is to make transit and active transportation the preferred transportation modes in order to reduce automobile dependency. The Transport Plan is composed of 21 strategic projects; nine of these are transit-oriented. The plan proposes several investments in active transportation and transit: extending metro and commuter rail lines, building several Light Rail Transit (LRT) lines in the center and a rail shuttle between downtown and the Montreal-Trudeau airport, implementing several Bus Rapid Transit (BRT) lines and new priority measures for buses, promoting car sharing, taxis and ridesharing and several measures to promote walking and cycling, including doubling the number of cycling paths in the next 7 years. It also has an important target: increase transit ridership by 8% by 2012, and 26% by 2021. The plan's overall strategy is a massive, systematic, and highly structured shift away from cars towards transit and active transportation modes.

Both the MTQ and the City of Montreal have adopted goals to reduce automobile dependency. Increasing transit is the favoured manner of attaining this goal. Although these goals have been adopted, they have not necessarily filtered through the planning process to the projects and plans being implemented. This is true not only of Quebec but generally of transportation plans in North America. Although transportation planning in general has moved towards new environmental and social goals, projects and policies still emphasize congestion relief (Handy, 2006a). The rebuilding of the Turcot Interchange is a prime example of a project that is still based on the idea of increasing mobility and reducing congestion.

### Modal Shift Strategy

Reducing automobile dependency through transit is based on the concept of modal shift. Rather than aiming to reduce the number of trips that are made, the goal of this strategy is to replace a percentage of solo automobile trips with transit, walking or cycling trips. For this shift to happen, transit must be more attractive than driving. In a study evaluating the elements drivers wanted improved in order for them to shift to transit for their daily work commute, respondents named frequency, reliability, convenient drop off sites, better connections and discount tickets as important factors. Security, more comfortable vehicles and better information were also mentioned but were less important (Kingham, Dickinson, & Copsey, 2001). The Agence métropolitaine de transport (AMT) for Montreal has placed modal transfer at the heart of its strategy to attain Kyoto objectives: reduce car trips by 25%, double the number of walking and cycling trips and double transit ridership over 15 years (AMT, 2003a).

The success of new transit projects relies on the application of incentives for development and disincentives for car users (Cervero, 1984). One example is to combine reserved bus lanes and high occupancy vehicle lanes with an aggressive parking policy (City of Montreal, 2008). In the proposal outlined below, transit improvements are combined with disincentives for car users in order to reduce the number of drivers commuting from the West Island to the center of Montreal. By reducing this number, it would be possible to redesign the Turcot Interchange to sustain this new transportation approach.

An interesting example of a city with a clear strategy to reduce car trips by replacing them with sustainable transport options is Beijing, China. In 2003, the City of Beijing decided to launch an ambitious plan to rectify the city's transit problems in time for the 2008 Olympics. The objectives were to improve the efficiency and attractiveness of urban transport. The target was that 60% of people's daily trips would be met by transit, 20 % by cycling, and the remainder by private automobiles (Beijing Transport Bureau, 2003).

Beijing has given public transit a strategic position in the city's sustainable development planning. According to the Beijing Public Transport Group, the city's transit system served 4.6 billion persons in 2007, a 13% increase from the previous year (Beijing 2008 Olympic Games). These transit improvements were combined with a restriction on driving which allowed cars with license plates ending in even numbers permission to drive one day and those with license plates ending in odd numbers the next. This resulted in a reduction of 1.82 million vehicles during the 20 days of the Olympic event, and a corresponding emissions reduction of 20% (Beijing 2008). Interestingly, after this restriction was lifted, 54% of car users in Beijing favoured its continuation, citing the perceived benefit of having clearer air and less congestion in a city which is consistently ranked among the world's worst in these categories (Beijing Airblog).

## The Turcot Today

The Turcot Interchange was built hastily in 1966-67, to be ready for Montreal's hosting of the 1967 World Exposition. It facilitated transport fluidity in Montreal by linking suburban areas to the downtown core. The interchange is a major highway junction between east-west traffic on the Highway A-20/Ville Marie axis and the north-south traffic on Highway A-15; it also allows direct highway access to downtown Montreal. Highway A-15 is the only North-South link between the industrial sectors Northwest of Montreal and the economic cradle of South-eastern Québec and the Northern New England States. From an economic perspective, therefore, its stability and reliability are crucial. (See **Figure 2.1a** in the colour section.)

The interchange, which accommodates over 300,000 vehicles per day, is situated in an urban residential area. The ramps and spans pass over residential parts of St.-Henri and Cote-St.-Paul, and connect neighbourhoods in Westmount and Notre-Dame-de-Grace (NDG).

The interchange is south of the St-Jacques Escarpment (a five-kilometre-long linear park), and north of the Lachine Canal. Directly to the west of the interchange are the Turcot Yards, once owned by the Canadian National Railway (CN) and now owned, since 2003, by the Provincial Government. Below the interchange are Notre-Dame and Pullman Streets, as well as a CN rail line, which passes in a tunnel underneath the interchange ramps. The interchange itself is made up of four elevated spans, the crisscrossing A-15 and A-20, and eight elevated ramps that connect the spans to one another. The structures are 20 to 30 metres high; originally they were raised to clear the CN rail line and boat traffic on the Lachine Canal, as well as to connect Highway A-15 at the St-Jacques Escarpment.

The need to rebuild the Turcot Interchange provides an ideal opportunity to begin constructing the sustainable city described in the Montreal Master Plan and Transport Plan. It is also an opportunity to reflect on the impact of a massive highway system in the middle of a sensitive and densely populated urban area and to encourage new strategies to meet Montrealers' mobility needs in a sustainable manner. The project should refurbish the junction in a way that reduces car and truck traffic, encourages public transit, and also enhances the quality of life of local residents.

**The MTQ Project**

The scenario adopted by the MTQ is currently under review by the Bureau d'audiences publiques sur l'environnement (BAPE), a government commission mandated to conduct public hearings on the environmental impact of the project. (See **Figure 2.1b** in the colour section.) Construction is slated to begin in November 2009. The MTQ based the design of the new interchange on certain requirements:

- Maintain current traffic capacity
- Improve road and highway access
- Maintain interchange functionality
- Improve safety and reduce congestion
- Decrease the number of elevated structures by 65%, to reduce maintenance cost and effort
- Integrate the highway with its surrounding environment
- Maintain traffic flow during construction

An important element in the project is the establishment of a "transportation corridor" along the base of the St-Jacques escarpment. This corridor would include all the CN rail lines, and Highway A-20, both running in East and West directions. It would permit a complete opening of the Turcot Yards for development, either of an urban, industrial, or recreational nature. Since the Turcot Interchange is only 6km from downtown Montreal this land could have significant development potential.

The project would put most of the elevated lanes onto embankments. Highway construction would take place beside or beneath existing structures. The Ville Marie Expressway would be expanded in width to 21 metres to create four 3.75-metre lanes in each direction, with 3 metre shoulders on each side.

**Health and Environmental Implications**

The MTQ project for the Turcot Interchange clearly does not meet the goals set out in its environmental policy. While it does not significantly increase the capacity of the interchange, it does not reduce it either and it includes no measures that favour transit. The City of Montreal has requested that the MTQ create reserved bus lanes on the east-west axis of the interchange; however, many believe this will be done by simply adding another lane, which would amount to increasing capacity (Bisson, 2008).

Furthermore, this project would have several serious health and environmental impacts. Transportation infrastructure is a significant factor in environmental degradation. Construction, rebuilding, maintenance, use, and even presence of infrastructure, particularly roads and railway lines, can have serious environmental consequences (Forman et al., 2003). Transport-related impacts are often cumulative: they combine with the effects of agriculture and industry to produce a magnified overall impact (Noble, 2006).

Highways in urban settings negatively impact their environment and pose serious health risks (see chapter 7 by Ferguson et al.). Both the construction and operational phases of a highway's existence affect the immediate environment, the regional environment, and local and regional residents' quality of life. Noise, air quality, socio-economic conditions, groundwater, soil and geology, and cultural and landscape assets are all potentially affected by the presence of a highway system such as the Turcot Interchange. Living in proximity to highways has measurable health-risks: in households within 100m of a highway cardio-respiratory illnesses can increase by 53%; within 200m of a highway, there is a 17% greater occurrence of babies born underweight (Drouin, 2008). These considerations are not addressed by the MTQ's proposal.

Furthermore, the scenario adopted by the MTQ will necessitate the partial expropriation of industries and residents. Presently the *Village des Tanneries*, a community of 450 residences, faces partial expropriation because of its proximity to the Eastbound Ville Marie Expressway.

The health hazards involved in putting highway lanes onto embankments in the middle of residential neighbourhoods are significant (Drouin, 2008; Thérien, 2008). Moreover, studies show that measures such as vegetation and noise barriers are not infallible mitigation methods. The results of a pollutant concentration study in Raleigh, North Carolina, for example, show that particulate matter engendered by highway vehicles may in fact occur in higher concentrations around noise barriers (Baldauf et al., 2008).

The shift of the railway lines to the north of their current position would also negatively impact the construction and operation of the new McGill University Health Centre (MUHC) hospital: noise levels, air quality, and accessibility to the hospital by emergency vehicles would all be affected. No mitigation measures are currently proposed to attenuate these impacts.

## A New Vision for Turcot

These health and environmental issues show that much more than con-necting highways is at stake. The goals stated in the City of Montreal's Master Plan and Transport Plan correspond to a collective long term vision for an urban development that shifts away from automobile dependency towards sustainable transportation modes and improves the quality of life of residents. These goals must not only find their way into the projects built on the island of Montreal, they must be their main focus.

Rather than maintaining traffic capacity, a project based on the sustain-ability goals enunciated in the MTQ and the City of Montreal's policy statements would aim to reduce it by favouring transit and by penalizing or increasing the cost of driving. Rather than expropriate residents and further the dismantling of a residential neighbourhood in the heart of the city, a sustainable transport project would aim to connect residents more efficiently with their destinations and promote active transportation.

The adoption of the Montreal Transport Plan 2008 presents a favour-able opportunity to reorient the Turcot project towards a transit-oriented design. In fact, the plan contains several transit projects that connect the West Island to downtown. These projects have in turn stimulated promot-ers to propose new transit initiatives specifically for commuters between these areas.

The implementation of four major transit initiatives, combined with a reconfiguration of the highway to reduce its impact on adjoining neigh-bourhoods, could significantly contribute to reducing the daily vehicle traffic on the highway and start Montreal on the process of lessening its dependence on the automobile.

### The Transit-Oriented Proposal

The transit-oriented initiative aims to reduce automobile dependence in Montreal. This proposal is composed of four parts that include: the imple-mentation of new transit projects, the elimination of ramps, the adoption of policies that discourage drive-alone trips, and the redesign of the inter-change to reduce highway capacity and enhance the quality of life of local residents.

#### GOALS

The primary goal of a sustainable, transit-oriented design for the Turcot Interchange is to reduce traffic volumes on the highway by getting people out of their cars and onto the transit system or the active transportation network. This concept directly informs the design of the project. The second goal is to ensure that the quality of life of local residents is preserved and enhanced. This includes avoiding both expropriation of neighbourhoods and creation of physical barriers which would further disconnect and isolate neighbourhoods from one another or from other parts of the city. The physical design of the interchange must have these considerations at heart.

OBJECTIVE

The immediate objective of the transit-oriented proposal is to reduce the vehicle traffic on the Ville Marie Expressway by 41% by 2016. This is equal to 68,000 vehicles per day (v.p.d.); going from the current volume of 163,000 v.p.d. to a volume of 95,000 v.p.d. This contrasts with the MTQ proposal, which would permit the highway's capacity to grow to nearly 200,000 v.p.d. by 2016 (Consortium SNC-Lavalin-CIMA 2008).

This objective is achieved by implementing four complementary initiatives: (1) improve transit to the West Island to remove 28,000 vehicles per day; (2) remove ramps and introduce new mass transit links from inner city neighbourhoods to the city centre thereby discouraging "short trips" using highway infrastructure to eliminate 40,000 vehicles per day; (3) introduce drive-alone disincentives such as parking controls and congestion pricing to consolidate these gains; and (4) redesign the Turcot Interchange.

1: INCREASE TRANSIT TO THE WEST ISLAND

At the forefront of this proposal is an expansion of public transit in the Highway A-20 corridor between downtown and the West Island. Four projects are currently under study for that corridor (see **Figure 2.3** in the colour section of the book): a rail shuttle connecting downtown and the Montreal-Trudeau Airport; an express tramway from Lachine and Dorval towards downtown Montreal; increased capacity on the Delson and Dorion commuter rail lines, and an additional service to Chateauguay; and the creation of reserved bus lanes on highway A-20.

All of these projects have the capacity of alleviating traffic; they must be part of a concerted strategy to reduce the number of vehicles that use Highway A-20 and the Turcot Interchange daily. Financing and building these transit projects should be an integral part of the Turcot project. Further details on each project follow below.

**Rail Shuttle.** The rail shuttle between downtown and the Montreal-Trudeau Trudeau Airport is one of the major transit initiatives adopted by the City of Montreal in the transport plan. It is currently under study by the AMT. Its objective is to accommodate 17% of airline travelers destined for downtown Montreal, a potential of 2 million travelers a year (City of Montreal, 2008).

**Lachine Tramway.** The Lachine tramway project is proposed jointly by Pabeco Inc. and the City of Lachine. The project proposes a commuter tramway linking the suburban area of Lachine with downtown. Lachine residents typically use the Highway A-20/Ville Marie Expressway as their commuter route (Bourque, Barrieau, & Lemire, 2007); implementing a tramway would specifically service this demographic. According to Pabeco, the Lachine Tramway would accommodate 30,000 passengers per day, representing 7.5 million users per year (Bourque, et al., 2007).

**Increased Rail Service.** The Dorion/Riguaud commuter rail line is currently operating over capacity. The AMT predicts that the population in the West Island and in the region directly west of the Island of Montreal will increase by 15-20% by 2026, as will the ridership of the trains for the home-work commute (AMT, 2006). Consequently, a new system and infrastructure is needed to meet the growing demand. It is vitally important that growing transport needs be met by an alternative to the automobile. As a response, the AMT and the provincial government have proposed increasing train frequency in this corridor by 35% (which corresponds to 230 more trains a week), and creating 10,000 new parking spaces in park and ride facilities around train stations. Increasing commuter rail ridership from the West into the downtown core will directly affect the number of cars using the east/west axis of the Turcot Interchange. It is expected 1.35 million additional passengers a year will use the trains (AMT, 2006).

**Reserved Bus Lanes.** The final element in increasing public transit to the West Island involves transforming one of the existing four lanes per direction on Highway A-20 to a reserved lane for transit and high occupancy vehicles. These reserved bus lanes would increase the efficiency of the numerous express bus lines that already use this circuit to bring commuters to the Lionel-Groulx metro station. Also, a transit service on reserved lanes would be faster since it would not be affected by traffic congestion. Reducing travel times will increase the appeal of transit and make it more competitive with the car. In fact, transit must offer comparable travel times with the automobile to conserve existing users and attract new ones (Vuchic, 2005).

Implementing each of these transit projects can potentially significantly reduce the number of vehicles using Highway A-20 and the Turcot Interchange to commute to and from the West Island. According to the 2003

Origin-Destination survey, 53,273 work trips to Montreal Center originate in the West Island during the morning peak alone (Enquête Origine-Destination 2003). Combined, these four transit projects have the potential of moving over 10 million people a year, or removing approximately 28,000 vehicles per day on the East-West axis of the interchange. The Lachine tramway and the improvements to the Dorion/Rigaud commuter rail line could move over 35,000 passengers a day, corresponding to 32% of the morning peak work trips between the West Island and Montreal Center. Reductions, whether at or exceeding these levels, or even at half these estimates, will alleviate the traffic on the interchange, which in turn will have a positive impact on pollution and health.

Transit projects can also have a significant impact on development around stations. They have consistently been shown to have a positive impact on land values, and when combined with appropriate planning and zoning can help to create denser neighbourhoods (Cervero & Duncan, 2001; Cervero & Landis, 1997; Chen, Rufolo, & Dueker, 1998; Knaap, Ding, & Hopkins, 2001).

## 2: REMOVE RAMPS AND INTRODUCE NEW TRANSIT LINKS

A second and more ambitious intervention is proposed for inner-city "short haul" traffic that uses the Turcot Interchange to access the Ville Marie Expressway. Through a series of strategic measures aimed at changing the behaviour of residents of Notre-Dame-de-Grace, Cote St. Luc, Hampstead and Verdun, we can further remove vehicles per day from the Expressway, theoretically as many as 40,000 vehicles each day.

The traffic reductions posited here are based on the traffic count volumes on the Turcot ramps provided by the MTQ in Atlas Quebec, http://transports.atlas.gouv.qc.ca/Infrastructures/InfrastructuresRoutier.asp (as of March 31, 2009). The reductions represent an indication, based on available data, of the scale of reductions that could be realised. More rigorous modeling, based on more accurate data, would likely lead to more robust estimates of traffic flows, driving behaviour and measures to reduce local use of the Turcot. In this sense, the numbers presented here represent goals which could theoretically be achieved by the replacement of solo automobile transport needs by public transit options similar to those described here.

A central element of this initiative would be the addition of express bus service lines that link the neighbourhoods of Notre-Dame-de-Grâce (NDG), Côte St. Luc, Hampstead and Verdun to the downtown core. Express bus lines would serve the neighbourhoods bordering the Decarie

Autoroute and would have exclusive access to the Ville Marie via the ramps connecting the highway to St-Jacques Street in Notre-Dame-de-Grace. Verdun express bus routes would access downtown via the new configuration of the Bonaventure highway.

The addition of these express bus service lines could result in a reduction in solo automobile use for back-and-forth traffic on the Ville Marie of 40,000 vehicles per day. This could be realized through, for example, the implementation of bus routes to specific ramps servicing the Ville Marie, as follows:

- The ramp linking the Decarie Autoroute to the Ville Marie, in the direction of downtown: this could potentially displace 11,000 v.p.d.
- The pair of ramps providing access to the Ville Marie from NDG would be reserved for emergency vehicles from the new MUHC hospital and public transit vehicles: this could potentially displace 21,000 v.p.d.
- Moreover, certain ramps could be closed entirely—such as those between the A-15 South and the Ville Marie—with 8,000 v.p.d. removed.

**Figure 2.2** portrays the projected reduction in traffic volumes from the express bus service.

## 3: DRIVE-ALONE DISINCENTIVES

The third part of the transit-oriented design aims at lessening automobile dependency by creating disincentives for car use. Implementing these measures will help to consolidate the reductions described above. By increasing the cost to the driver, these strategies make driving less attractive and make transit more competitive with the car. Additionally, these measures can be used to finance transportation improvements and projects, including transit, which is another step towards sustainability.

Two such measures are proposed by the City of Montreal in their Transport Plan 2008 and can be implemented immediately as part of a concerted transit-oriented strategy.

**Parking Control.** The City of Montreal proposes to reduce parking spaces in the downtown core in order to discourage car use (City of Montreal, 2004). Parking availability plays an important role in the choice of driving versus transit; by reducing the number of available parking spaces, the city is actually favouring transit over the car (Weiss, 2004). Also, by ensuring drivers pay for parking, car use will be reduced and revenue will be raised for public transit initiatives through the user-pay principle (Gouvernement du Québec, 2007). This initiative could have a substantial

**Figure 2.2** Traffic counts for the Ville Marie Expressway, actual and projected

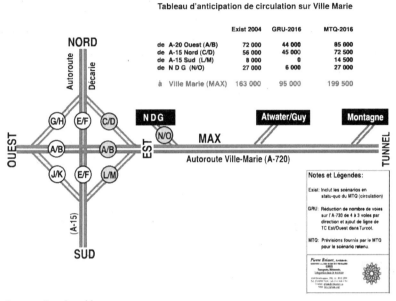

Tableau d'anticipation de circulation sur Ville Marie

|  | Exist 2004 | GRU-2016 | MTQ-2016 |
|---|---|---|---|
| de A-20 Ouest (A/B) | 72 000 | 44 000 | 85 000 |
| de A-15 Nord (C/D) | 56 000 | 45 000 | 72 500 |
| de A-15 Sud (L/M) | 8 000 | 0 | 14 500 |
| de N D G (N/O) | 27 000 | 6 000 | 27 000 |
| à Ville Marie (MAX) | 163 000 | 95 000 | 199 500 |

Interpreting the table:
- The column "Exist 2004" shows the daily numbers of cars that currently use the four ramps (volumes provided by the MTQ in Atlas Quebec, see endnotes),
- The column "GRU-2016" shows the potential daily numbers of cars that could be found on these ramps after the reductions specified by initiatives 1 and 2 above (28,000 v.p.d. from West Island reductions, 40,000 v.p.d. from inner core neighbourhood reductions).
- The column "MTQ-2016" shows the MTQ projections of daily traffic counts for the MTQ's Turcot Interchange reconstruction alternative (data from Consortium SNC Lavalin-CIMA 2008).

Source: Traffic count data retrieved from Atlas Quebec and Consortium SNC Lavalin-CIMA 2008.

impact on traffic volumes. A study in San-Francisco found that a $3 'parking cashout' caused 45% of the reduction in drive-alone trips predicted by the transportation model; a cashout means offering employees who receive free parking the choice of using the space or receiving the equivalent value in cash (Lewis, 1998).

Furthermore, both minimum and maximum parking requirement policies would be implemented by the City to encourage the use of public transit. Adding maximum parking requirements for certain land uses is an important measure since having only minimum parking requirements subsidizes driving and inflates parking demand (Manville & Shoup, 2005). **Congestion Pricing.** The City of Montreal is considering placing tolls on the roads surrounding Montreal: the goal is to discourage car use, reduce negative impacts of automobiles, and increase city revenues. These tolls could be an efficient way to increase revenue as well as control the number of cars that access the city center. Tolls would be electronic and would not disrupt traffic flow. Payment would vary between times of the day; evenings and weekends would be free; and tariffs would depend on weather or air quality conditions, such as smog and snowstorms. Buses, taxis and emergency vehicles would be exempt. These tolls could raise between $425–$450 million per year, which would be dedicated to the implementation and operation of public transit (Lavallée, 2008). Other cities around the world have implemented similar congestion pricing systems to reduce traffic. London, Singapore and Stockholm, for example, report 15% to 45% reductions in traffic after implementing these systems (Environmental Defense Fund, 2007).

By combining increased transit options with disincentives for drive-alone trips and for car commutes to the city center, it could be possible to reduce the number of vehicles using the Turcot Interchange daily. This would not impact the number of trips between downtown and the West Island; it would simply move people out of cars and into sustainable transport modes. Simply by implementing the transit projects contained in the Montreal Transport Plan and those planned by the City of Lachine and the AMT, combined with parking policies and congestion pricing, it would be possible to reduce the number of cars going from the West Island to downtown by over 28,000 vehicles daily.

Finally, with additional improved transit links from adjacent neighbourhoods that also use the Ville Marie Expressway, such as Notre-Dame-de-Grace, Cote-St Luc, Hampstead and Verdun, we believe it is possible to remove an additional 40,000 vehicles per day, for a combined reduction of 68,000 v.p.d. (See **Figure 2.2** for projected traffic counts on ramps that feed into the Ville Marie Expressway).

In light of the above observations, it is possible to imagine a different design for the Turcot Interchange. This design would support the new transit projects and would impact the health and quality life of nearby residents in a positive manner. It could reconnect rather than disconnect neighbourhoods and free up valuable land for development to create sustainable, transit-oriented neighbourhoods. It would also discourage the use of the highway for short trips between neighbourhoods.

4: REDESIGNING THE INTERCHANGE

Rather than simply rebuilding the Turcot Interchange in the same manner and for the same purpose, there is an opportunity to redesign it in a fashion that would still fulfill its transportation function, while placing transit and people first.

The City of Montreal is currently building the first phase of a plan to replace the Bonaventure Highway with an urban arterial network, linking downtown to the Champlain Bridge through a series of urban streets and boulevards (Société du Havre, 2007). This approach should also be applied to the Turcot Interchange. Some ramps currently encourage expressway use for travel between adjacent neighbourhoods, something which is incommensurate with the goals of sustainability and quality of life. Replacing this access by an urban arterial boulevard on Atwater Street would solve this problem, and also eliminate the higher-emissions option of travelling the extra 5 km of highway to get from downtown to the Champlain Bridge.

Certain inner-city public transit initiatives could also serve to reduce traffic on the ramps of the interchange: for example, adding express bus lines from adjoining boroughs to downtown and designating ramps exclusively for their use could potentially reduce traffic by 40,000 vehicles per day, as we have seen above.

The Turcot Rail Yards, abandoned for the last five years, could be reinstated for parking, maintenance, and repair of trains. The Yards are in a naturally occurring physical depression between the St-Jacques Escarpment and the escarpment on the South side of the Lachine Canal, and tend to accumulate smog and air pollution. As such, they are unfit for residential development but ideal for use pertaining to the maintenance of the public transit initiatives (Brisset, 2008). At a minimum, given its industrial history, these hectares would require decontamination at considerable cost.

The St.-Jacques Escarpment would be protected from disturbance and encroachment. It could be developed as a public park and a natural sound barrier which would enhance the neighbourhood (see chapter 4, by Asch).

Finally, when possible, whatever is rebuilt must remain within the existing right of way and a serious debate should be held as to the pertinence of replacing this urban expressway (see chapter 3 by Sijkpes for the renovation option).

## Reducing Traffic Volume

There is a strong relationship between the reduction in the number of cars and the reduction in pollution. Studies have shown that higher traffic count data indicate higher environmental impacts, while reduced traffic volumes indicate a corresponding reduction in impacts. Traffic reduction has a positive effect on health and safety, air quality, noise environment quality, greenhouse gas emissions, soil quality, hydrological system health, and urban micro-climate integrity (Canter, 1996; Forman, et al., 2003; Morris & Therivel, 2001; Noble, 2006).

The proposed transit projects, parking policies and congestion pricing have the potential of reducing car traffic by over 68,000 vehicles per day. By reconnecting neighbourhoods and eliminating inter-neighbourhood traffic, the redesign of the interchange would create more sustainable neighbourhoods where walking and cycling are favoured by residents. It is entirely within reach to reduce car traffic and improve health and quality of life.

All the above mentioned projects are presently in the preliminary study phases. Concerted political will and leadership are now required to integrate them within a large scale transportation project.

Now we must turn our attention to more ambitious projects.

★ ★ ★

## Planning For The Future

Reconstructing the Turcot Interchange is a short term project with long term impacts. The MTQ project maintains the status quo, projecting the current situation onto future generations. The transit-oriented proposal uses the Turcot project as an opportunity to begin building a large scale transit network which improves neighbourhoods. Turcot is only the first step in what can be a major transformation of Montrealers' commute and travel behaviour, and also of the city center's urban landscape.

Once the new transit projects that connect downtown with the West Island are implemented, it would be possible to begin other phases to increase transit access and decrease car use.

The Ville Marie Expressway brought with its construction the massive expropriation of neighbourhoods. It is still a major barrier in Montreal's center. Conversely, transit projects can be catalysts for development (Gospodini, 2005) and provide an opportunity to develop new neighbourhoods.

Transit projects attract people and businesses and have a positive impact on land values (Cervero, 1984; Cervero & Duncan, 2001; Chen, et al., 1998; Knaap, et al., 2001).

A long term vision for Montreal, therefore, would see the Ville Marie Expressway dismantled or buried in central neighbourhoods and new transit projects built to ensure Montrealers' transportation needs are met in a sustainable manner. The current proposal suggests several steps for the achievement of these goals, the first of which is the implementation of a new high-capacity transit line: the Metro-Express.

### The Metro-Express

In order to reduce car use in the downtown area, the number of lanes on the Ville Marie Expressway should be reduced. Two lanes in each direction could be reserved for transit vehicles allowing installation of a new major transit project. Because several bus lines and metro lines are currently at capacity (City of Montreal, 2008), a third high-capacity transit line could be built to connect the east and west of central Montreal to downtown. A Light Rail Transit (LRT) line running from the Rosemont borough through downtown to Montreal West and NDG would reduce passenger congestion, supplement tramway services, and further discourage automobile use (see **Figure 2.4** in the colour section).

This LRT could be built using only existing infrastructure, such as CP rights-of-way, and the Ville Marie tunnel. It should be an express train, stopping less frequently than the metro. It should be designed to go underground and on the surface in order to permit future expansions on the network.

### Reclaim Central Neighbourhoods

Another long term goal should be the removal of the Ville Marie Expressway between St-Remi, at the heart of the Turcot Interchange, and Atwater on the Eastern border of Westmount. Removal of this section would permit the restoration of an urban neighbourhood devastated by the demolitions engendered in the early 1970s, and would heal an urban wound which has isolated the area since.

In order to do this, the number of vehicles using the highway must be reduced and the ridership of transit increased: it must be feasible to dramatically reduce car access to the city center or to split the remaining volume up, slow it down and redirect it into the existing city street network. This depends on planning strategically to attain this goal, rather than following existing trends.

The ultimate goal is to completely remove the Ville Marie Expressway between the Turcot Yards and Atwater Street, and subsequently convert the Turcot Interchange into a three-way junction. The highway would stop before the city center; cars would travel on the arterial network through the center; or preferably people would use transit, walk or cycle to get around the center. A 2-lane high-speed inner-city slipway, accessible from Atwater Street and terminating at Papineau, would still permit emergency vehicle traffic to travel towards the east and the west. The area that is freed up by the removal of the Ville Marie could then be re-greened or developed in a way which is no longer inhibited by the presence of an expressway. The resulting free land would have increased value due to its proximity to the downtown core (see **Figure 2.5** in the colour section).

## Conclusion

The need to rebuild the Turcot Interchange presents a unique opportunity to redesign it according to sustainable long-term goals. Rather than maintain the status quo, the transit-oriented proposal uses the Turcot project as an opportunity to begin building a large-scale transit network and improve neighbourhoods.

Several major transit projects connecting the West Island to the city center are currently under study: the rail shuttle connecting downtown to the Montreal-Trudeau airport, a Tramway between Lachine and downtown, increased service on commuter rail lines and reserved bus lanes combined with new inner-city BRT lines. These transit projects have the potential of reducing car traffic by over 68,000 vehicles per day. By enforcing the parking policies and congestion pricing proposed by the City of Montreal, this approach would assure this decrease, instead of increasing the volume by 36,500 vehicles per day as proposed by the MTQ (Consortium SNC Lavalin-CIMA 2008). It would then be possible to redesign the interchange to reconnect neighbourhoods and eliminate inter-neighbourhood traffic, creating more sustainable neighbourhoods where walking and cycling are favoured by residents. By implementing these projects as part of a concerted transit-oriented strategy for the Turcot Interchange it is entirely within reach to reduce car traffic and improve health and quality of life of residents. The Turcot project is only the first step in what could be a major transformation in Montrealers' commuting and traveling patterns, and in the city center's urban landscape.

## Acknowledgment

The authors would like to thank Stephanie Titman and Catherine Doucet for their contribution to preliminary versions of this paper. Credit should · be given to Assumpta Cerdá for editing and conducting additional research. The authors would also like to acknowledge the help of Jason Prince, Dr. Jochen Jaeger, Dr. Craig Townsend, Daniel Bouchard, and Philippe Coté for their help and comments in reviewing this paper. All opinions and remaining errors in this paper are the responsibility of the authors.

## References

AMT (2003a). *Kyoto, une stratégie en transport des personnes pour la région métropolitaine de Montréal*: Agence Métropolitaine de Transport.

AMT (2003b). *Portrait des transports collectifs dans la région de métropolitaine de Montréal*: Agence Métropolitaine de Transport.

AMT (2006). *Rapport d'activités: Rapport annuel de l'Agence Métropolitaine de Transport*: Agence métropolitaine de transport.

Baldauf, R., Thoma, E., Khlystov, A., Isakov, V., Bowker, G., Long, T., et al. (2008). Impacts of noise barriers on near-road air quality. *Atmospheric Environment, 42*(32), 7502-7507.

Beijing 2008 Olympic Games. Retrieved 02/5/2009, from http://beijingolympic2008.wordpress.com/2008/01/08/beijing-public-transportation-volume-increased/

Beijing 2008. The official Website of the Beijing 2008 Olympics Retrieved 02/05/2009, from http://en.beijing2008.cn/spirit/beijing2008

Beijing Airblog. Retrieved November 2008, from www.beijingairblog.com

Beijing Transport Bureau (2003). Development of the public transport system in Beijing.

Bourque, P., Barrieau, P., & Lemire, A. (2007). *Mémoire de l'arrondissement de Lachine : pour la relance du tramway vers Lachine*: PABECO Inc.

Brisset, P. (2008). *Reconfiguration de l'Échangeur Turcot : Analyse de sols et des fondations dans la cour Turcot*. Montréal: Groupe de Recherche Urbaine.

Brisson, B. (2008). Échangeur Turcot: une diminution de la circulation réclamée. *La Presse*.

Canter, L. W. (1996). *Environmental Impact Assessment 2nd Ed*. Singapore, McGraw-Hill.

Carmona, M., & Sieh, L. (2008). Performance measurement in planning towards a holistic view. *Environment & Planning C, 26*(428-454).

Cervero, R. (1984). Light rail transit and urban development. *Journal of the American Planning Association*, 133-147.

Cervero, R., & Duncan, M. (2001). Transit's value-added: Effects of light and commuter rail services on commercial land values.

Cervero, R., & Landis, J. (1997). Twenty years of the Bay Area rapid transit system land use and development impacts. *Transportation Review, 31*(4), 309-333.

Chen, H., Rufolo, A., & Dueker, K. (1998). Measuring the impact of light rail systems on single-family home values: A hedonic approach with geographic information system application. *Transportation Research Record*(1617), 38-43.

City of Montréal (2003). *Plan de transport. Portrait et diagnostic. Mobilité des personnes et des biens.*

City of Montréal (2004). *Master Plan.* Montréal City of Montréal.

City of Montréal (2008). *Plan de Transport* Montréal: City of Montréal.

Consortium SNC-Lavalin/CIMA (October 29, 2008). *Projet de reconstruction du Complexe Turcot; Impacts sonores* Rapport sectoriel, Annexe H : données de circulation (djme) prévues en 2016 avec le projet de reconstruction du complexe turcot. Prepared for the Quebec Ministry of Transport.

Drouin, L. (2008). *Transport et santé publique : Enjeux et solutions.* Paper presented at the Colloque Écosanté, ACFAS.

Drouin, L., Morency, P., King, N., Thérien, F., Gosselin, C., & Lapierre, L. (2006). Impacts sanitaires de la circulation automobile. *Routes & Tranports, 35*(3).

DSP (2006). *Collective Urban Transportation, a Question of Health. Annual Report on the Health of the Population .* Direction de la Santé Publique, Agence de Santé Publique et de Services Sociaux du Québec.

Enquête Origine-Destination (2003). *Mobilité des personnes dans la région de Montréal, version 03.a période automne.*

Environmental Defense Fund (2007). Congestion Pricing: A smart solution for reducing traffic in urban centers and busy corridors. Retrieved from http://www.edf.org/page.cfm?tagID=6241

Forman, R. T. T., Sperling, D., Bissonette, J. A., Clevenger, A. P., Cutshall, C. D., Dale, V. H., et al. (2003). *Road Ecology. Science and Solutions. .* Washington, DC.: Island Press,.

Gospodini, A. (2005). Urban development, redevelopment and regeneration encouraged by transport infrastructure projects:The case study of 12 European cities. *European Planning Studies, 13*(7), 1083-1111.

Gouvernement du Québec (2007). *A Collective Commitment: Government Sustainable Development Strategy*: Ministry of Sustainable Development, Environment and Parks.

Handy, S. (2006a). *New planning processes for new transportation planning.* Paper presented at the AESOP Conference on Transportation Planning, a Policy Design Challenge? .

Handy, S. (2006b). *New Planning Processes for New Transportation Policies* Paper presented at the AESOP Conference on Transportation Planning a Policy Design Challenge? , Amsterdam.

Kingham, S., Dickinson, J., & Copsey, S. (2001). Travelling to work: will people move out of their cars. *Transport Policy, 8*(2), 151-160.

Knaap, G. J., Ding, C., & Hopkins, L. D. (2001). Do plans matter? The effects of light rail plans on land values in station areas. *Journal of Planning Education and Research*(21), 32-39.

Lavallee, A. (2008). *Le péage régional : une tarification socialement responsable dédiée au financement du transport collectif.*

Lewis, S. L. (1998). Land use and transportation: Envisioning regional sustainability. *Transport Policy*, 147-161.

Manville, M., & Shoup, D. (2005). People, parking, and cities. *Journal of Urban Planning and Development, 131*(4), 233-245.

Morris, P., & Therivel, R. (2001). *Methods of Environmental Impact Assessment, 2nd Ed.* London, New York.

MTQ (1994). *Éléments de problématique et fondements de la politique sur l'environnement du Ministère des Transports du Québec.*

MTQ (2006). *Québec Public Transit Policy: Better Choices for Citizens.*

Noble, B. F. (2006). *Introduction to Environmental Impact Assessment: A Guide to Principles and Practice.* Canada: Oxford University Press

Société du Havre (2007). *Transformation of the Bonaventure Expressway at the downtown gateway from Saint-Jacques Street to Brennan Street* Montréal.

Thérien, F. (2008). *Health Impact of the Turcot Reconstruction.* Paper presented at the Coalition of Community Groups Mobilized around the Transport Québec Turcot Project.

Vuchic, V. R. (2005). *Urban Transit: Operations, Planning, and Economics.* Hoboken, NJ: John Wiley and Sons.

Weiss, S. (2004). *Le controle de l'offre de stationnement comme outil de regulation de la mobilite: l'exemple de la region metropolitaine de Montréal. Universite de Paris VIII -St Denis.*

**CHAPTER THREE**

# What is Still very Good at Turcot

*Pieter Sijpkes*

In this chapter alternatives to the two main parts of the plan for the Turcot Inter-change and Turcot Yards put forward by the Ministère des transports du Québec (MTQ) will be outlined. These counter-proposals were developed this winter at McGill University when my second-year architecture students and I looked into the opportunities and problems related to the Interchange and the Yards. The chapter is divided into four sections: a summary; a counter proposal for the repair (rather than the proposed demolition) of the Turcot Interchange; a counter proposal to modify the MTQ's plans to reduce the Falaise St Jacques to a highway embankment; and a final section in which all Montrealers are encouraged to visit and survey the Turcot area in person. For those who prefer to stay in their armchairs, I provide you with a short walking tour, in words, of my own recent visit to this extraordinary place.

## Summary

This stage in the course of events surrounding the Turcot Yards project can be compared to the time when things went wrong with the Apollo 13 mission. The pivotal moment in that mission came when its crew was asked by ground control to focus on what was still good on the ship rather than on what went wrong. We are at a similar pivotal point in the 40-year saga of the Turcot Interchange and Turcot Yards, and so I ask "What is still good at Turcot?"

The MTQ project is informed by two major decisions, both of which should be reviewed from an Apollo mission perspective. The first decision is that the many spans of the Turcot Interchange (and the associated d'Agrignon and de la Verendrye sections) should be torn down and replaced by "lowered" structures.

I argue that a repair-in-place option should be revisited, particularly for the crucial North-South axis of the highway. The East-West axis can be secured for now, and eventually dismantled, as detailed in another chapter of this book (chapter 2 (by Brisset and Moorman)). Repair-in-place of the viaducts would not require expropriation of housing units, would preserve the many hectares of useful land below the structures, and would eliminate the permanent scarring and dividing of neighbourhoods with embankments.

The second MTQ decision is to move highway 20 and the CN tracks towards the Falaise St. Jacques. This plan is reckless and expensive. The highway and the tracks are fine where they are (straight and on grade) and how they are (in good condition). There is no need to "open up" at great expense 100 hectares of developable land that are already opened up. Instead, the Falaise St. Jacques could be widened into a real and unique linear park, and it and part of the restored St. Pierre River could be protected from highway noise by a simple landscaped berm. The remaining land could then be used for what it is well suited—train shunting and maintenance, and truck-rail transfer activities. (This would have the additional humanly beneficial effect of making neighbourhoods like Point St. Charles and Cote St. Luc less noisy as such rail activities were moved away from them.)

Unlike the single plan presented by the MTQ, the similar Alaskan Way Viaduct case in Seattle has for years seen experts and the general public alike rigorously debate as many as ten options. Importantly, a repair-in-place option was recently given new life with the commission of a $150,000 independent study from the prestigious engineering firm T.Y. Lin. Regardless of the outcome of this effort, it shows a commendable openness to other points of view.

The lives of the Apollo 13 crew were saved by focusing on what was still good on their craft; the environment and the pocketbooks of Montrealers will no doubt benefit from a similar change in focus in the MTQ Turcot strategy.

### Counterproposal 1: The ups and downs of the Turcot Interchange: a repair option outline

The construction of the Turcot Interchange coincides with my arrival in Montreal in 1966. I saw the concrete being poured; I rode across the interchange in a borrowed car the day it opened. And there was a lot more fresh concrete curing in Montreal at that time. Luigi Nervi's Place Victoria had just been declared the highest concrete high-rise in the world, Ray Affleck's Place Bonaventure, the first real mega-building on earth, was being denuded of its formwork, and Moishe Safdie's boxes of Habitat '67 were just settling in on top of each other.

The Turcot Interchange looming high above the black landscape on its hundred-foot stilts, lit by two parallel, continuous bands of built-in fluorescent lights, fit very well into this brave new world. However, the "alien landing lights" were gone after two years (aluminum wiring and aluminum casing do not last long when splashed with salt water in winter). And the concrete of the interchange soon showed signs of distress.

Reinforced concrete, when soaked for prolonged periods in water (salt or not) will absorb it to a certain depth. If this depth is more than the thickness of concrete covering the reinforcing bars (two or three inches), these bars start to rust. When steel rusts it expands in volume by about ten times. Once water gets in, "rust inflation" will soon push away concrete covering it, exposing yet more steel to the corrosive influence of salt water. ("Spalling" is the technical term for this process).

The Turcot Interchange was very tightly designed—there are no shoulders, so snow cannot be pushed to the side to clear the road, and lots of road salt has been used to keep the easily frozen, exposed spans navigable. The brine resulting from this process has flowed like a river down the sloping roadways, not easily finding drains. Expansion joints (planned "cracks" in the structure to allow it to expand and contract) have formed virtual waterfalls, and the water has also found its way into many unplanned cracks in the structure. The deck of the Interchange (the upper section that constitutes the bed of the roadway) is made up of eight feet deep by eight feet wide, hollow "caissons." In some places as many as seven caissons are found side by side. When finally inspected through newly cut access-holes, some of these hollow caissons were found to contain significant amounts of standing salt water which had infiltrated the hollow structures. If you set out to weaken a concrete structure, this is the easiest way to do it.

Over the years, many efforts have been made to repair flaws in the Turcot Interchange. The built-in lights were replaced by standard pole-mounted lights; the spalling concrete was patched up over and over again. The caissons were drained and the roadways above were repaired many times. But the expansion joints remained problematic and the spalling of the structure became so pervasive in recent years that the concrete was no longer patched; instead, in many places and for all to see, it was covered with two layers of mesh (one coarse layer covering a fine one) to prevent lumps of spalled concrete from falling onto people and cars passing underneath.

After the collapse of the de la Concorde overpass in Laval a few years ago, worries about the structural integrity of the spans brought about almost desperate measures. Anchors were installed from the top of some of the caissons to the bottom, in the hope of preventing structural disintegration and possible collapse of the caissons (particularly the outer ones, and mostly at the expansion joints). The Interchange is monitored 24 hours a day now, at a cost of several million dollars a year; it is not uncommon to see worksites set up where emergency structural repairs are quickly made.

**Figure 3.2** Elevation of a typical Turcot span after repair. Graphic by Stuart Kinmond

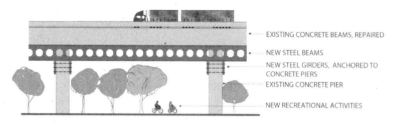

EXISTING CONCRETE BEAMS, REPAIRED

NEW STEEL BEAMS

NEW STEEL GIRDERS, ANCHORED TO CONCRETE PIERS

EXISTING CONCRETE PIER

NEW RECREATIONAL ACTIVITIES

Yet, I think the Turcot Interchange should be fixed rather than torn down. Mine is a minority voice. The engineering community has decided that tearing the structure down and replacing it with a system of on-grade access ramps and a few overpasses will solve "the problem" the same way the Pine-Park interchange problem was solved. But a replacement operation will be complicated, costly, and damaging to the environment.

As with a heart-lung operation, the system has to be kept fully functioning while the replacement operation takes place. The volume of concrete debris created would be huge, and as many as 200 apartments would have to be demolished, which would squeeze the last breath from the small enclaves left after the massive demolitions required by the initial construction. The free space now existing underneath the overpasses will be cut up and made useless by the earth embankments. (For a preview of what is planned, the Anjou interchange is offered as a model, located in the North-East part of Montreal.) Finally, the estimated cost of the whole operation stands at 1.5 billion dollars. I am always wary of figures like 1.5 billion. Why not 1.6 or 1.4? Because nobody really knows what the final cost will be! Look at the Notre Dame East reconstruction. Two years ago it was estimated to cost 750 million dollars. Last month, without a spade having touched the ground, the cost has gone up to 1.5 billion dollars. Need I also point again to the sad saga of the Super-hospitals?

(Perhaps we ought to be concerned with ballooning costs for highways eroding limited public funds, funds that ought to be directed at mass transit infrastructure.)

There are two schools of thought in construction: the conservative school (of which I am a member) and the more radical slash-and-burn school. I use the word "conservative" here as it was used during Apollo 13, the movie. When faced with possible disaster, Apollo 13's mission control director Gene Kranz said, "What do we got on the spacecraft that's good?"

**Figure 3.3** Cross section of part of the Turcot Interchange

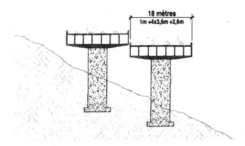

This attitude helped save the lives of the astronauts. Taking a comparable stance toward fixing the Turcot will surely save a lot of grief and an incredible amount of money. Again, the "slash and burn" approach to hospital renewal (now over ten years in limbo) comes to mind, and it is interesting to see how reality has forced the McGill University Health Centre to include a renovated Montreal General Hospital building in its original tabula rasa plan. But I digress.

There are many precedents for major infrastructure projects that have followed the conservative "modify rather than demolish" route (see endnotes). An important one for all to see is barely a mile away from Turcot and goes back over a century. The original Victoria Tubular Bridge was built in 1859 in the form of a single wrought iron tube on 24 ice-breaking piers. A single train track ran through the tube. Towards the end of the century this arrangement was woefully inadequate, and a new bridge was needed. By reusing the 24 piers, and using the steel tube as a scaffold, the current truss-structure came into being. According to the *Montreal Herald* of the time:

**Figure 3.4**

While the old bridge, entire, weighed 9,044 tons, the new bridge weighs 22,000 tons. The total length of the bridge is 6,592 feet; number of piers, 24; number of spans, 25; length of central span, 330 feet; length of side spans, 242 feet.

While the width of the old bridge was 16 feet, the width of the new bridge will be 65 feet. The height of the old bridge was 18 feet; the height of the new bridge over all is 28 feet.

The total cost of the new bridge, which provides double tracks for railroad trains, driveways for vehicles on each side, and footwalks on the outside of the driveways, is about $2,000,000. The contract price of the old Victoria Bridge was $6,813,000.

$6,813,000 for the original 1859 bridge, and $2,000,000 for the 1897 one. It seems that re-use certainly paid this time! In the link section at the end of this chapter there are references to several other large infrastructure repair and reuse examples.

## Renovate the Turcot

My proposal for the Interchange consists of three ideas. One, public safety has to be put first and foremost; a collapse like the the de la Concorde overpass must at all cost be prevented. The solution is to reinforce all or most of the spans of the (until now) self-supporting concrete deck by deep steel beams. A triage should be done, to identify spans that need extra support right now, those that can wait, and those that don't need reinforcement. (This process is common in earthquake country: at a Tokyo train station viaduct I recently counted four different structural reinforcement systems within five spans; the motto there clearly was: repair if you can at all costs.) Fortunately installing supporting beams is quite easy in most places because the Interchange is elevated above grade as much as 100 feet, leaving ample room for a new, prefabricated structural beam (or even an arch in the long span cases) to be inserted, without obstructing passage below. Once the original spans have this additional support, there is no longer any worry about structural collapse even if part of the caisson structure has deteriorated.

It is important to note that the de la Concorde collapse was due to an unusual "lip joint" that was badly designed, badly executed, and not maintained for 40 years. The Turcot, on the other hand, is a continuous, indeterminate structural system; this system has a lot of redundancy in it, which, combined with an extra support system underneath, will guarantee ample structural strength for many years.

If added measures have to be taken to bring the structure up-to-date for earthquake resistance, this is the time to do it. In my references below I give links to many examples of retro-fit earthquake proofing.

Two, the roadways on the deck and the expansion joints will have to be redone. This process should ensure excellent drainage is achieved to minimize water infiltration. In addition, the minimum amount of salt and the maximum amount of harmless grit should be used in winter from now on to keep the roads navigable. It is said that it is impossible to have waterproof expansion joints, but every time I drive over them on Rene Levesque in front of Place Ville Marie, I realize that there is no water leakage in the underground city below. If it can be done there it can be done

**Figure 3.5** Model and section of repaired Turcot span (cross-bracing with castellated beams not shown in the section for clarity)

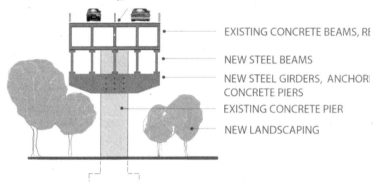

EXISTING CONCRETE BEAMS, RE

NEW STEEL BEAMS

NEW STEEL GIRDERS, ANCHOR
CONCRETE PIERS

EXISTING CONCRETE PIER

NEW LANDSCAPING

**Figure 3.6** Detail of the existing concrete structure supported from below by a new grid of man-high steel beams.

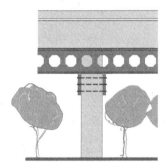

at the Turcot. The spalled surface concrete now so evident should also be repaired using the latest techniques.

Third, the maintenance of the structure should be rigorous and continuous. Following the practice of Gothic Cathedrals (where maintenance structures were built into the fabric at the time of construction), the new under-deck support structures that I suggest would serve as platforms for future maintenance of the concrete deck without the need for expensive scaffolding every time a flaw becomes apparent. The steel girders portrayed in the illustrations are of the castellated type, which feature large openings in the web, allowing easy access by maintenance personnel.

It is my view that taken together these three measures will allow the interchange to continue to function for many years to come. During this time there will be opportunities to improve and extend the public transit system with projects like the rapid rail line to Trudeau airport, thus relieving some of the pressure on the Interchange. Cars, on average, have become smaller and lighter since the Turcot was built, so the rather tight dimensions of the roadways are actually becoming less onerous as time passes.

Finally, one great advantage of the methodology sketched above is that it can be completed in steps, starting tomorrow. These steps can be small and we can learn from them as we go along; costs can be tightly controlled during a step-by-step repair of the system; and no interruption of traffic and no displacement of people will be required.

**Counterproposal 2: A case for the expansion of the Falaise St. Jacques into a linear park.**

The MTQ's plans call for the realignment of the CN tracks and Autoroute 20 close to the Falaise St. Jacques. Public pressure against this move has recently led them to change their original plans: the location of the right-of-way for the highway and the tracks are now set back from the Falaise by 15 metres rather than being right up against it. This change in alignment makes the cost of relocation even more onerous when calculated as a charge per hectare of land. My student group has focused their efforts on the design of the expanded Parc Falaise St. Jacques.

One intriguing idea brought up by them was that the Park could accommodate functions that are not available in any of the other Montreal parks, such as horse riding, mountain biking and orienteering. One student proposed in some detail a steel cable suspended from a tower in Terry Fox Park (a small park at the top of the Falaise on St. Jacques Street) that would allow visitors to slide down to a pavilion at the bottom, across where rue Pullman now runs and where she had restored the St. Pierre River. It was wonderful to see these ideas bubble up, and it reminded me of the time in the early 70's, where, as students, my study mate and I proposed a green park along the banks of the Lachine Canal (including a bike path) to incredulous critics.

Terry Fox Park provides an excellent location for the construction of an access pavilion towards the new "Parc Falaise St. Jacques."

*People of Montreal! Get to know your Turcot Yards and Interchange! Go visit and imagine what could be...*

★ ★ ★

## Getting to know the Turcot

Who has not heard: "Traffic is heavy going in from the Turcot to the Ville Marie this morning." or "Coming from the West on the 20, avoid a stalled car in the right lane heading onto the Turcot." or " Smooth sailing through the Turcot this morning."

Everybody in Montreal knows about the Turcot. Whether you're young or old, fit or not, a car driver or a bicyclist, French- or English-speaking, you can't avoid hearing the endless morning and evening traffic reports that clutter Montreal's airwaves. The nearly 300,000 cars that daily negotiate the many tentacles of the Interchange apparently form a large enough market to make this interminable Turcot traffic intelligence broadcast-worthy round the clock.

But does everybody really know the Turcot? Not really. In fact many people in Montreal know the lay of the land in Old Orchard Beach or Fort Lauderdale better than the in and outs of the Turcot Interchange and Yards. And no wonder, because even today, with the property in Government hands, anyone entering the domain is technically considered a trespasser. Before the site was acquired by the MTQ it was railway property, tightly patrolled and strictly off-limits.

Yet the Turcot Interchange and the Turcot Yards and the Falaise St. Jacques all belong to we taxpayers. That is, to you and me, and whatever plans may be implemented by the MTQ will in effect be financed by us. So, to better inform yourself, let me suggest that you leave these words and take a trip to the Turcot Interchange to have a look 'round for yourself. After all, taking a peek at your own property can hardly be seen as a real infraction. My nine architecture students and I made such a trip in February through deep snow (and have been back several times) as part of a Turcot-focused McGill architecture design course. These trips have been a revelation. I provide links to the blogs of several intrepid explorers who have paved the way over the last few years, in my endnotes.

### A walking tour of the Falaise St Jacques and the Turcot Yards

You can choose the high road by starting from the Vendome Metro, or take the low from the St Henri Metro, using the map composed here from Navigateur Urbain (a great online resource that for free lets you combine photos and plans and allows you to measure distances quite precisely—a vital tool in judging a project this size!)

Let me give you some tour-guide direction as you make a loop starting from on high at Vendome and ending below at St Henri. Head west along de Maisonneuve and go left at Decarie (don't miss the view at the still-vacant site of McGill's Superhospital at your left, a silent reminder of the way grand ideas get bogged down, their cost spiralling out of sight). Continue to the end of Decarie Boulevard at St. Jacques, and turn left; cross the street to continue at the South side of the street, which gives you a top-down view of the interchange, always animated by streams of cars, "rubies coming and diamonds going" as the song says.

Continue over the massive Decarie Expressway viaduct towards Girouard, (enjoy the view from here to the South, over the Lachine Canal, all the way to the St. Lawrence and Mount Johnson beyond) and keep going till you get to Terry Fox Park, a green strip shoehorned between St. Jacques and the Falaise St. Jacques below to your left. Take the gravel path that meanders down the middle of the park until you're about halfway through, where you will see a hole in the fence (obviously well-used) separating the park from the relatively steep slope below. (This is the location of the student-proposed entrance pavilion shown above. From this pavilion you would be able to access a belvedere and a set of stairs down the slope).

Slip through the hole, and enter a strange green wonderland. In old photographs the Falaise appears as a treeless dump, but in several campaigns dating back to the reigns of Jean Drapeau and Jean Doré, the site was gradually cleared of the most noxious detritus, trees and bushes were planted, and it is now a green oasis rooted between the rocky outcrops of stone and concrete fill that were left in place after the clean-up.

Slip-slide down the first part of the slope that is particularly steep, and enjoy a break at the level, overgrown pathway used by intrepid NDG strollers and dog walkers. After your break, continue down the slope till you reach the bottom. You have now descended about 100 feet. Take care, because it is often wet down at the bottom. Jump over the creek and get back to level, terra firma on the asphalt of Pullman Road.

You can now see the green Falaise from the bottom up, and, scanning counter-clockwise, you see the three-kilometre extent of Rue Pullman stretch empty before you like a country road. (In the student plans this road would be turned into a meandering bike and pedestrian path that would run the full length of the park. It would intertwine with the restored St. Pierre River.)

Keep on turning, and the equally empty Turcot Yards stretch out in front of you as far as the eye can see. Turn a bit more and Autoroute 20 comes into view in the South, about 400 metres away. (This view would be changed by my proposal for a landscaped berm that would silence the traffic noise from Autoroute 20 and the CN tracks.)

Finally, turning some more, the amazing outline of the Turcot Interchange appears. Curved, intertwining concrete ribbons on high stilts form a strange, groaning monument of enormous proportions. Its great height and the complete emptiness of the land below is an important aspect of the Interchange because it allows the easy insertion of support structures from below without compromising headroom, as I discussed above.

Now head East on Pullman toward the interchange and pass underneath it to walk among the almost 100-foot-high portals created by the overpasses You can't help but notice the lower noise level down here. The wonderful panels of graffiti adorning the massive piers of the viaducts seem fitting here. You can imagine mature trees growing in this amazing landscape, which we might call le forêt Turcot. A gate crosses the way from the Yards but there is ample room to slip through.

Pullman ends at St. Remi. Make a right turn there and duck underneath the beginning of the Ville Marie expressway above. Go on a few feet until you get to rue Cazelais on your left. In almost every window on the street an "Action Turcot" poster is on view, and understandably so, because almost 200 housing units in this pleasant looking neighbourhood will be demolished if the MTQ plans are executed. This community has been in the forefront of mobilizing scrutiny of the MTQ plans.

Rue Cazelais abuts Rue Desnoyers. Make a left there and a right when you get back to Rue St. Jacques. St Henri Metro station is six blocks down the road. You have made the Turcot loop! And you have seen great views from the top, accomplished the descent of the green Falaise, surveyed the vast empty Yards, marvelled at the height and size of the Interchange, and walked by, maybe talked to, the people of the threatened neighbourhoods.

### In Conclusion

It is important that we Montrealers realize what is at stake here and also important that we realize the now eerily empty Yards encompass an area more than twice the size of Old Montreal. We should know that the Falaise could be a linear park almost four kilometres long. Now a unique green area in the heart of the city, it could be a recreational resource to rival sections of the Lachine Canal paths in popularity. Finally, clearly under-standing that the massive Intersection structure is still doing the job it was designed to do a mere forty years ago, we must look seriously into upgrading it instead of simply demolishing it.

The fate of this huge "packet of urbanity" has been, lock stock and barrel, in the hands of the Ministère des transports du Québec, who have made, in isolation, major decisions. In my view these decisions are mis-guided, and, at the very least should be subjected to serious scrutiny. Add-ing some poetry and a good dose of fiscal restraint to the review decision matrix would be a good start.

### References

There are some interesting blogs on the Turcot Interchange and Yards that are continuously upgraded. Andy Riga's excellent blog has many links to other Turcot sites embedded in it. See Riga, Andy. "Turcot: A Feature in Five Blog Postings." *The Gazette* (April 17 2009), http://communities.can-ada.com/montrealgazette/blogs/metropolitannews/archive/2009/04/17/turcot-a-feature-in-four-blog-postings.aspx.

### Websites

Below are some of the websites that students and I created during the McGill 2009 winter term. Some of the student work: http://www.arch.mcgill.ca/prof/sijpkes/turcot_book/

The website below is a short version of the original two-pronged al-ternative proposal to the MTQ plan: Sijpkes, Pieter, and McGill University U2 architecture students. "Two Counter-Proposals to Plans Now under Consideration by Le Ministere Des Transports Du Quebec." (April 2009), http://www.arch.mcgill.ca/prof/sijpkes/Turcot/intro.html.

The website below has many chapters that look at other examples of re-use and repair of infrastructure: Sijpkes, Pieter. "Pour Turcot Autrement, Yes!" (2008), http://www.arch.mcgill.ca/prof/sijpkes/U2-winter-2008/presentation-turcot/cover.html.

A particularly interesting link is to the Victoria bridge project in Roches-ter, UK, where an arch bridge is turned into a truss bridge without inter-rupting traffic: Gibson, James. "Construction Gallery." The Rochester Bridge Trust, http://www.rbt.org.uk/bridges/oldgall.htm.

The many twists and turns of Seattle's decade-long Alaskan Way Via-duct saga can be traced on-line in posts too many to list.

# Turcot, un modèle mondial ou un clou dans le cercueil pour le verdissement et la diversité biologique du grand Sud-Ouest?

*Patrick Asch*

Les villes et les arrondissements autour du complexe Turcot comptent très peu de milieux naturels.

La falaise Saint-Jacques, en superficie un des plus grands espaces verts de la ville de Montreal, et rivale du Parc Mont-Royal, pourrait devenir l'épine dorsale d'une vaste trame verte dans le Sud-Ouest de Montréal, liant Montréal-Ouest, Notre-Dame-de-Grâce, le Sud-Ouest et Westmount. Ce parc linéaire, réinventé, et le site du Campus Glen du CUSM, pourraient ensemble former une vaste toile verte. Le nouveau parc est ainsi lié aux autres espaces verts dans un vaste réseau d'espaces verts et de pistes cyclables, rassemblant quelque 500 hectares d'espaces verts.

Cependant, dans la version du projet Turcot présentée par Transport Québec en 2009, aucune intervention de naturalisation des milieux n'est prévue tandis que l'autoroute et les chemins de fer seront accolés à la falaise sans barrière sonore. Avec une zone tampon de seulement 15 mètres, le bruit rebondissant sur la falaise sera si fort que la faune et les utilisateurs y seront sans cesse dérangés. Ce parc en devenir sera alors condamné à l'abandon tant par la faune que par la majorité de la population.

Ce chapitre explore les options pour une approche verte dans le secteur du Turcot.

Les villes et les arrondissements autour du complexe Turcot comptent très peu de milieux naturels. En fait, leurs communautés ont accès à moins de milieux naturels que probablement toute autre population au Québec. Dans ces secteurs, plusieurs passent des mois, voire des années, sans sortir des zones urbaines. Certains peuvent même passer des vies entières sans fréquenter un marais, un bois ou un cours d'eau.

Pourtant, quelques siècles auparavant, les environs du complexe Turcot comportaient une diversité biologique parmi les plus importantes du Québec. Tout autour, on retrouvait des érablières à caryers, des prucheraies et des pinèdes. Dans la cour Turcot actuelle, ancienne cour de triage CN, on pouvait rencontrer une rivière, un lac et des milieux humides où castors, orignaux, ours et autres animaux typiques de ces endroits trouvaient à l'époque refuge. Les sols de la région comptaient parmi les plus riches de la province, attirant ainsi les nombreux colonisateurs européens. C'est alors que fut entamé un processus qui transforma ces sols en terres agricoles, puis en villages et finalement en métropole.

Le remaniement du complexe Turcot touchera de très grandes surfaces dont les quelques 100 ha du secteur de l'ancienne cour Turcot longeant l'autoroute du Souvenir entre les échangeurs Saint-Pierre et Turcot. Une importante portion de ces terrains sera libérée une fois la reconstruction terminée et pourrait servir à la reconstitution de milieux naturels. La reconstruction du Complexe Turcot pourrait du même coup avoir un impact négatif sur un talus boisé long de plus de 3 km au nord de l'autoroute du Souvenir appelé la falaise Saint-Jacques. Cependant, cette falaise—considérée comme un écoterritoire et à protéger par la Ville de Montréal—est actuellement inaccessible du haut du talus étant donné la présence de commerces sur presque toute sa longueur.

## Turcot, une opportunité

Plusieurs organismes communautaires et environnementaux du milieu voient, dans le réaménagement du complexe Turcot, une opportunité de verdissement et de naturalisation. Alors que les milieux naturels et que les espaces verts du secteur forment présentement de petits îlots disparates, il serait possible de les relier pour former une trame verte qui longerait le grand Sud-Ouest. Cette trame, formée de corridors verts qui permettraient tant à la faune qu'aux humains de circuler d'un milieu à l'autre, pourrait rassembler jusqu'à 500 ha d'espaces verts dont 250 ha seraient constitués de milieux naturels. Nous formerions ainsi, au sein d'un des secteurs les plus dépourvus d'espaces verts et de milieux naturels au Québec, une trame verte de la taille des plus grands parcs de l'agglomération de Montréal. (La **figure 4.1** est reproduite dans la section en couleurs de l'ouvrage.)

Nous ne trouvons donc pas seulement de l'espace pour reconstituer des milieux naturels et un écoterritoire qui pourrait devenir accessible à la suite de l'aménagement du Complexe Turcot, nous proposons aussi une mise en valeur qui permettrait de maximiser les potentiels récréatifs et écologiques de même que la qualité de vie et l'image de marque des communautés environnantes.

Cependant, dans la version du projet Turcot présentée par Transport Québec en 2009, aucune intervention de naturalisation des milieux n'est prévue tandis que l'autoroute et les chemins de fers seront accolés à la falaise sans barrière sonore.

Avec une zone tampon de seulement 15 mètres, le bruit rebondissant sur la falaise sera si fort que la faune et les utilisateurs seront considérablement dérangés. Ce parc en devenir sera alors condamné à l'abandon tant par la faune que par la majorité de la population.

Ainsi, alors que la reconstruction du complexe Turcot représente une opportunité majeure permettant la création d'une trame verte qui augmenterait la diversité biologique du secteur et l'accessibilité communautaire à des espaces verts, la vision actuellement proposée rend les milieux existants quasi-inutilisables en plus de former un obstacle qui couperait la trame en petits morceaux isolés. Dès lors, une question se pose : serait-il possible de modifier le plan d'aménagement du complexe Turcot afin de rendre possibles ces interventions?

Deux problématiques potentielles doivent toutefois être considérées : le besoin d'accoler l'autoroute sur la falaise et la possibilité de créer les liens (voir : pistes cyclables et voies piétonnes) autour du secteur.

D'abord, rien ne semble justifier que l'autoroute soit déplacée vers le nord pour l'accoler à la falaise. Au contraire, tout semble compromettre ce projet. Étant donné la présence d'un ancien lac et de divers milieux humides, le sol le plus près de la falaise est, juste sous sa surface, complètement saturé d'eau. Cette eau souterraine rendrait la construction d'une autoroute plus coûteuse puisqu'on devrait auparavant y reconstruire une base stable. En revanche, nous retrouvons, à partir de l'ancien chemin de fer du Grand Tronc, vers le sud de la cour Turcot, une base incomparablement plus stable, fruit de travaux effectués il y a quelque 150 ans. Partant, construire dans le secteur sud entraînerait des économies majeures.

### Les bénéfices d'un Turcot intégrant verdissement et diversité biologique

Quant à la possibilité de construire des liens, plusieurs types sont faisables. Pour le transport actif, des pistes multifonctionnelles sur les échangeurs, ou passant sous une section d'autoroute, seraient imaginables. Pour la faune, la présence de milieux boisés de part et d'autre de l'autoroute pourrait permettre au moins à diverses espèces d'oiseaux de voyager de bord en bord. Combiné à la création d'un pont vert végétalisé—un concept que la ville de Montréal explore actuellement pour le secteur du ruisseau Demontigny adjacent à l'autoroute 25—le projet permettrait aussi à cette faune de circuler sans difficulté même avec l'autoroute. Par le fait même, ce pont vert pourrait comprendre une piste multifonctionnelle permettant à la population de le traverser. Il est donc à la fois possible, et de déplacer l'autoroute, et de créer des liens permettant à la faune et aux humains de circuler du secteur vers les autres sections de la trame verte du grand Sud-Ouest.

Une telle approche permettrait nombreux autres bénéfices tant au niveau du design du secteur qu'au niveau de la population. En ce qui a trait au design du complexe Turcot, il faut ajouter qu'il sera nécessaire de gérer les eaux pluviales du secteur. L'aménagement considéré pourrait viser le rassemblement de ces eaux afin de former des milieux humides qui pourraient les filtrer naturellement. Ces bassins ralentiraient du même coup le drainage des eaux vers les égouts et favoriseraient une évaporation naturelle, réduisant à la fois l'impact des pluies provenant de l'ensemble des quelque 100 ha de la cour Turcot sur le système d'égouts et les coûts associés au développement d'infrastructures gérant ces eaux.

Mais il est aussi possible d'envisager une autre solution. Transport Québec pourrait choisir de maintenir l'échangeur Turcot surélevé afin de former, sous l'autoroute même, un lac qui serait alimenté à la fois par les eaux pluviales et par une déviation d'eau en provenance du canal de Lachine. Ce lac, situé non loin du centre récréatif Gadbois, un centre au dessous des bretelles de la structure Turcot, pourrait jouer un rôle tant écologique que récréatif. Par exemple, le personnel du centre Gadbois pourrait, pendant la période estivale, s'occuper d'une plage et, pendant la période hivernale, s'occuper de patinage ou de pêche blanche et même entretenir un réseau de pistes de ski serpentant vers la falaise Saint-Jacques.

Pour la population, les bénéfices de ce milieu naturel irait toutefois bien au-delà de la gestion de l'eau ou de quelques loisirs ayant cours autour d'un lac. On améliore la qualité de l'air alors que le Sud-Ouest, selon le département de santé publique de Montréal, comporte des problèmes de pollution de l'air significatifs. On réduit l'effet d'îlot de chaleur alors que les quartiers du Sud-Ouest sont fortement minéralisés. Tandis que le complexe Turcot apportera une pollution significative et un effet d'îlot de chaleur, planter une quantité significative de végétaux serait une solution logique qui réduirait, au moins en partie, ces problèmes et contribuerait à améliorer la santé des communautés environnantes. De plus, les milieux naturels constituent des lieux propices à d'autres loisirs dont la marche et le vélo. Les activités qu'on pourrait y faire contribueraient ainsi à améliorer davantage la santé de la population.

Mais cette communauté a besoin d'emplois. Chaque année, selon le Gouvernement américain, 23% des Américains, soit 66,1 millions des citoyens, dépensent annuellement quelque 1449 $US uniquement à observer la faune. Plus près de nous, selon le Gouvernement du Québec, 3,4 millions de Québécois génèrent 2,9 milliards $ en investissement économique et

32 000 emplois en effectuant des loisirs en nature. Le tourisme associé aux milieux naturels est en fait le domaine connaissant la plus grande expansion dans la province. Cependant, ces dépenses ne sont pas effectuées uniquement en régions éloignées. Au contraire, selon Aventure Écotourisme Québec, elles sont en majorité effectuées près des milieux urbains car les résidents—mais aussi les touristes—n'ont en général que quelques heures ou que quelques jours à allouer à leurs loisirs. En somme, s'il y a des milieux urbains prêtant à la marche, au vélo, à l'observation d'oiseaux ou à tout autre loisir en plein air, alors tant les résidents du secteur que les touristes seraient susceptibles de visiter les environs du complexe Turcot. Les milieux naturels créeront ainsi de l'emploi et injecteront de l'argent au sein de l'économie locale. Cet impact économique est si important dans les milieux urbains que plusieurs études rapportent que ces attraits deviennent un point de vente pour les agents immobiliers. D'autres études évoqueront d'une augmentation de la valeur foncière de 5 à 32% en bordure de tels milieux.

En somme, le simple fait de déplacer la construction de l'autoroute au niveau de la cour Turcot et de prévoir des liens formant une trame verte ouvre la porte à diverses possibilités de naturalisation et de mise en valeur d'espaces verts qui auront des impacts majeurs non seulement sur la qualité de l'air, les effets d'îlot de chaleur, la santé des populations, l'accessibilité à des lieux de loisirs, la diversité biologique et la création d'une trame verte, mais aussi sur l'investissement local, la disponibilité d'emplois, la valeur foncière et l'image de marque des quartiers environnants. Ajoutons à cette énumération la possibilité de maintenir une structure surélevée au niveau de l'échangeur Turcot. Il devient alors envisageable d'intégrer la création d'un lac avec marais filtrant à cette trame verte qui, en plus d'accroître tous ces impacts, aura par le fait même l'avantage de réduire le fardeau des eaux pluviales sur le système d'égout montréalais.

### Québec et la protection de son territoire : quelques chiffres

Le Gouvernement du Québec, tout comme les communautés environnant Montréal, pourrait du même coup sortir gagnant de cette entreprise si l'autoroute était maintenue au sud de l'ancien chemin de fer du Grand Tronc. Le 29 mars 2009, le premier Ministre Jean Charest annonçait que Québec avait protégé plus de 8% de la superficie de la province en tant que milieux naturels et se donnait pour objectif d'en protéger non moins de 12% pour 2015. Or, quand nous examinons ces données en profondeur, on découvre que sur les 135 326 km² protégés par le Gouvernement du

Québec, 8 km² seulement (c'est-à-dire 0.006% du total) sont protégés sur l'ensemble de la superficie de la CMM.

Si Québec désire être équitable face à la disponibilité des milieux naturels autour de Montréal et si Québec veut profiter au maximum des nombreux impacts financiers associés aux milieux naturels, alors Québec devra tenter de se rapprocher du 12% de milieux naturels protégés autour de Montréal. Présentement, avec seulement 8 km² protégés, Québec n'a atteint que 1,7% de son objectif de 12%, ou 461 km², au niveau de la CMM. La valeur des terrains étant très élevée autour de Montréal, Québec devra profiter au maximum de toute opportunité de protéger les milieux naturels qui n'impliquera pas d'investissement en argent.

Si Québec maintenait l'autoroute au sud de l'ancien Grand Tronc, alors il serait possible d'utiliser la surface de 55 ou de 60 hectares entre l'ancien Grand Tronc et la falaise Saint-Jacques pour des fins de protection d'espaces verts et de naturalisation de milieux dégradés, et ce, sans investir un sou, mais plutôt en échangeant les terrains avec un développeur qui s'engagerait à effectuer un projet de verdissement en bordure de la falaise Saint-Jacques.

### Une proposition innovatrice : Meadowbrook

Par exemple, il se trouve, à environ 1,5 kilomètres de la falaise Saint-Jacques, sur la continuité de cette falaise, vers l'ouest, un espace appelé le terrain Meadowbrook qui sert présentement de terrain de golf. Il fut établi sur des anciennes terres agricoles, il y a 90 ans, selon des techniques qui, à l'époque, modifiaient peu le sous-sol. Le terrain est bordé de boisés possédant des vestiges de forêts locales datant de plus d'un siècle. On y découvre également un des derniers ruisseaux de Montréal, un étang, des plaines inondables, un sol n'ayant pas été modifié en profondeur depuis la colonisation et un potentiel majeur pour l'arrêt des oiseaux migrateurs. Son sous-sol inaltéré possède de plus un potentiel archéologique majeur, car il pourrait abriter des vestiges de peuples amérindiens ayant habité, il y a plus de 8000 ans, les rives de l'ancien lac Lampsilis, dont l'eau atteignait alors le haut de la falaise Saint-Jacques.

Ce site de 57 ha—le plus facile à naturaliser de tout le grand Sud-Ouest de Montréal, grace à son usage historique (il n'est aucunement pollué)—, est présentement menacé de développement. Le développeur, afin de tenter d'obtenir l'approbation pour son projet, propose de construire de l'habitation en effectuant un projet qui serait l'exemple à suivre en Amérique du Nord en développement durable tout en naturalisant par la suite 50% de sa superficie. Il n'en demeure pas moins que, malgré un effort louable de construction selon les principes de développement durable, il détruit le plus intéressant milieu à naturaliser de l'ensemble du grand Sud-Ouest.

Une solution qui serait intéressante tant pour Québec que pour le développeur, de même que pour les résidents des environs, serait de proposer à celui-ci un échange de terrains : 55 ou 60 ha entre le vieux Grand Tronc et la falaise contre 57 ha au terrain Meadowbrook, sous condition que le 50% de verdissement prévu par le développeur crée une zone tampon en bordure de la falaise et que le développement soit positionné afin de réduire le plus possible le bruit de l'autoroute dans le nouveau parc. Le développeur bénéficierait alors d'un terrain plus accessible qu'auparavant, Montréal bénéficierait d'un projet de développement durable unique qui servirait d'exemple aux autres villes du Continent et Québec aurait protégé plus de 80 hectares de nouveaux milieux qui, selon les normes de l'Union internationale pour la conservation de la nature (UICN), correspondrait à la catégorie 4 des milieux naturels protégés. Finalement, uniquement autour de Meadowbrook, de la falaise Saint-Jacques et de la cour Turcot, la communauté bénéficierait de près de 100 ha de milieux naturalisés pouvant s'intégrer des plus facilement à la trame verte du grand Sud-Ouest.

Le réaménagement du complexe Turcot peut donc, soit empêcher le verdissement à la grandeur du grand Sud-Ouest, soit devenir, sans coûter plus cher, un modèle mondial quant à l'intégration du verdissement et de la diversification biologique à un projet autoroutier. La proposition est donc déposée et nous ne pouvons qu'espérer que le Gouvernement du Québec choisisse d'en profiter.

CHAPTER FIVE

# What Sort of Problem is the Replanning of the Turcot Interchange?

*Raphaël Fischler*

No less than five levels or institutions of government are involved in urban planning for Montréal and its region: the province of Québec, the Communauté métropolitaine de Montréal, the Agglomération de Montréal (and surrounding regional county municipalities), the Ville de Montréal and other municipalities, and the boroughs of the city of Montréal. Of particular importance in this context is the fact that the provincial Ministère des Transports is in charge of important road projects that will have major impacts on local development and quality of life in Montréal neighbourhoods. These projects should be seen as urban projects which are responsive to local needs and conditions, and not only as transportation projects which facilitate the flow of people and goods on a regional or national scale. The institutional clash between levels of government is augmented by conflict among professionals from different fields and between officials and residents. All these tensions are reflected in debates over the redesign of the Turcot Interchange. This experience, among others, suggests that better collaboration is needed between the province and the city, between engineers and planners, and between experts and lay people. The "problem" of the Turcot Interchange is therefore a technical problem, an urban problem, and a political problem.

Transportation planning, like all urban planning in Montréal, is complicated by the diversity of actors, in particular state actors, involved in the process. Multiple levels of government participate in decision-making on the issue: on the Island of Montréal alone, the federal government has authority over bridges, the province of Québec is in charge of highways, the agglomeration of Montréal is responsible for major local roads, and the boroughs and reconstituted municipalities deal with local streets. In addition, multiple stakeholders from the private sector and from civil society weigh in on public decisions: producers of trains and buses, engineering firms and construction companies push their wares and services, while environmental organisations, trade associations and neighbourhood groups work to "sell" their views.

For example, when the province decided to extend highway 25 from Montréal to Laval, the city of Montréal found itself on a collision course with the province (and with the city of Laval, which pushed for the project). For Québec officials and for businesses owners in the vicinity of the new link, completion of the highway network will facilitate economic development in the northeastern quadrant of the metropolitan region. For Montréal officials and for environmentalists, it will first and foremost foster sprawl in eastern Laval and on the North Shore. For local residents, it will bring increased noise and pollution and will require proper mitigation.

In a superficial way, all parties in the debate agree that transportation planning ought to promote sustainable development. Beyond the common use of the expression, there is in fact little agreement on what ought to be done to improve mobility and accessibility. Should the highway network be completed on the South Shore in order to lessen truck traffic on the island? Should the metro network be grown to service new areas of Montréal? Should the bicycle network be expanded to allow for travel by bicycle during the six months of warmer weather in the year? What projects should be given priority for use of scarce public resources?

These questions are at the heart of planning. Planning is the identification of goals and the mobilisation of resources to attain them. In other words, it is the selection of priorities for the expenditure of time and money. Yet many plans, especially comprehensive or master plans, present general goals, long wish lists and precious few priorities. The first Master Plan of post-merger Montréal earned an award of excellence from the Canadian Institute of Planners (Ville de Montréal 2004a). But it contains so many objectives that mayors will be able to pick and choose projects within that framework for the next few decades and do so according to the direction of prevailing political winds.

Likewise, the Transportation Plan of Montréal is made up of 21 projects among which no clear priorities are set (Ville de Montréal 2007). All 21 projects are worthy, and it is not clear, from reading the document, how their respective worth and, hence, their order of implementation, might be established. This vagueness is a virtue in a turbulent environment. Politicians may claim that an official plan represents "a social contract" with the community (Ville de Montréal 2004b), but the twin needs to hold numerous forces in balance at all times and to ensure reelection periodically make them weary of contracts that might bind them prematurely to certain decisions. From a politician's point of view, the freedom of choice afforded by most master plans is a strength rather than a weakness.

Disagreement on spending priorities is inherent to democracy. It is especially intense in our complex system of decision-making, with actors taking their cues from different constituencies and setting their aims at various geographic scales. But the problem is not resolved once priority projects have been selected for implementation. For every chosen project, differences of opinion are equally intense with respect to the specifics of design. Even if there were a broad consensus on the idea that Rue Notre-Dame Est ought to be modernised to facilitate truck traffic to and from the Port of Montréal, there would be no agreement on how, exactly, this ought to be done. Should the new roadway separate truck traffic from other traffic? Should it, rather, include separate lanes for public buses? Or should it be combined with a right-of-way for a light-rail line? If the project is limited to the redesign of some intersections to make truck-flows smoother and therefore less productive in diesel fumes, should Notre-Dame be put below ground while cross-streets remain at grade, or should the opposite be done?

For higher-level government planners, efficiency in the flow of people and goods is the main criterion of evaluation to select the right solution. For local residents, who have organized themselves in a "coalition to humanize Notre-Dame street" (*la coalition pour humaniser la rue Notre-Dame*), quality-of-life issues such as noise and pollution, together with access to the shoreline, are the greatest concerns. For municipal planners, the best possible balance between efficiency and quality of life is the objective. The list of questions that transportation planners and urban designers will have to answer is very long, and each question on that list can have multiple, contradictory answers. It is not surprising, therefore, that the Notre-Dame project has been on the drawing boards for several years now.

★ ★ ★

When considering such cases as Rue Notre-Dame and the Turcot Interchange, it is perhaps worth keeping in mind that the design of roadways has a long history of conflict in the twentieth century. Much of modern urban planning can be seen as a battle over the proper place of the automobile in the city. Clarence Perry's Neighbourhood Unit Idea is a response to the increasing dangers that car traffic posed to children on their way to school, the park or the playground (Perry 1929), and the Radburn Idea of Clarence Stein and Henry Wright presents their way of keeping a proper balance between car and foot traffic in residential environments (Stein 1957). Le Corbusier's answer to the question of how to deal with the growing car traffic was a radical separation of modes: walking would take place on pathways going through park-like settings and driving would be done on highway-like thoroughfares (Le Corbusier 1987 [1925]).

Jane Jacobs wrote her classic *Death and Life of Great American Cities* largely as a rebuttal of Le Corbusier work; whereas he condemned the street as an obsolete urban form, she rehabilitated it as a key ingredient of urban livability (Jacobs 1961). Jacobs was an arch-enemy of Robert Moses, who, like other modernists of his time, plowed urban highways through old urban fabric to make downtowns, many of them struggling in the face of rapid suburbanization, more accessible by car (Fogelson 2001).

Lewis Mumford (with Benton McKay) had, as early as the 1920s, called for the design of highways strictly as inter-city links, free of urban sprawl (the "townless highway") and kept up the fight after WWII for taming the automobile in the city and making it one transportation mode among others (Mumford 1964 [1958]).

Despite the warnings and recommendations of such writers and designers, city after city, including Montreal, succumbed to the brutal charm of urban highway construction (Lortie 2004). Thus was born the trench of the Ville-Marie expressway, whose injury to the city is only now being healed, and that of the Décarie expressway. Thus came into existence the smaller interchange at avenue du Parc and avenue des Pins, which was turned back, just a few years ago, into an intersection at grade, and the much, much larger interchange of highways 20, 15 and 720, which is now the object of heated discussion. This debate pits government officials against residents, provincial authorities against local bodies, and engineers against urban planners and designers. This is not new: highway design has for a long time been an arena of conflict between different world-views of the city, of transportation and of the role of experts and lay people in decision-making (Ellis 1996).

The *Plan d'urbanisme* and the *Plan de transport* of Montréal seem to have put this debate to rest (Ville de Montréal 2004a, 2007). Some road segments, such as Boulevard Cavendish, need to be built in order to open some inaccessible "brownfield" sites to residential and commercial redevelopment, some axes of commercial transport, such as Rue Notre Dame, must be modernized, and the transportation of goods by truck and by train must be rationalised. But otherwise, priority is to be given to collective or active modes of transportation: the train, the metro, the tramway, the bus, the shared car, the bicycle, the feet.

The Tremblay administration seems to have made a clear choice: from now on, transportation planning will serve not only the interests of car and truck drivers; it will help to "integrate other fundamental dimensions: environmental preservation, air quality, quality of life, neighbourhood tranquility and ambiance, the safety and health of citizens, the quality, comfort and design of public spaces, social equity and the sums to be invested" (Ville de Montréal 2009; author's translation).

This will not eliminate controversy, though, because important choices still remain to be made among the projects to be implemented. For instance, should we pour massive amounts of money into a new tramway system when so much still needs to be done to repair, upgrade and expand our existing systems (metro, bus, roads)? Conflict over transportation choices will continue, too, because some of the most important transportation projects on the island of Montréal are under the control of the province.

★   ★   ★

The controversy over the Turcot Interchange offers a good illustration of the sort of complex, conflict-ridden process by which decisions are being made, and are likely to continue being made, with respect to Montréal's transportation infrastructure. Here, as in the cases of Autoroute 25 and of Rue Notre-Dame, the main protagonists are the province, the city and local residents. My purpose here is not to analyse in detail the respective positions of each party. Rather, it is to stress the fact that we are dealing with a multi-layered conflict: one that pits different levels of government against each other in a battle over planning jurisdiction, that opposes different professions in a debate over the proper aims and forms of urban planning, and that confronts experts and lay people over the distribution of costs and benefits.

The formal responsibility for rebuilding the Turcot Interchange lies with the Ministère des Transports (MTQ). The MTQ is formally committed to fostering sustainable transportation (*promotion de la mobilité durable*), to protecting the natural and built environment (*mise en valeur du patrimoine écologique, culturel et social*), and to ensuring the proper integration of its projects into their context (*intégration de la route à son milieu*) (MTQ 2009; author's translation). Its stated vision of transportation is comprehensive: the ministry's plans reflect "an integrated vision of transportation" in which every transportation mode is considered in order to "answer the mobility needs of people and goods" and in which transportation planning is coordinated with land-use planning, environmental planning and social-economic development planning (*ibid.*).[1]

To many, these statements do not reflect the behaviour of the ministry in specific projects. Perhaps the gap between lofty words and concrete actions is implicitly acknowledged in the ministry's mission statement: "The mission of the ministère des Transports du Québec is to ensure the mobility of people and goods throughout Québec on safe, efficient transportation systems that contribute to the sustainable development of Québec" *(ibid.).*[2] The focus on mobility and on efficiency reflect the traditional paradigm of transportation engineering in which speed of movement and economy of means are supreme values. The conflict over the Turcot project is in large part a disagreement between provincial planners and local residents; but to a certain extent it is also a clash of values that is internal to the ministry itself, perhaps even internal to individuals in that agency.

Old habits die hard. Senior professionals, who have decades of work behind them, have been trained at a time when automobile transportation received priority over other modes and when speed was given precedence over other objectives. It is not easy for such persons to shift perspective and to recognize the limitation, let alone the faults, of their beliefs. The same holds for urban planners and designers schooled in Modernism, for whom the old had to make way for the new and who were told at some point that historic preservation trumps affordable housing production (see for example Forester, Fischler and Shmueli, 2001: chapter 9).

The inertia of individual thinking is easily complemented by institutional conservatism to produce stasis in the behavior of large organisations. Official policy changes therefore do not lead quickly or easily to changes in behaviour at the level of the project. Although the Québec government already declared its allegiance to the doctrine of sustainable development in the late 1970's (Léonard 1978), its decisions have continued to reflect the status quo in many instances.

Declarations on "integrated planning" and "planning for integration" notwithstanding, the search for mobility and efficiency remain high on the agenda of transportation planners, and for good reasons, too. It is imperative, for economic reasons, that Montréal and its region be endowed with an efficient, well-maintained infrastructure system that enables trains, trucks, buses and automobiles to transport workers and products with ease, safety and speed to their destinations. It is not narrow-mindedness that pushes planners of the MTQ to recommend that Rue Notre-Dame be made more convenient for truck traffic to and from the Port of Montréal, a major source of wealth and employment for Montrealers. At the same time, it is not antagonism to economic development that makes critics of the MTQ claim that a more comprehensive plan should put more emphasis on transport by rail or that a revised design for the road should divide the costs and benefits of the new axis somewhat differently among regional and local stakeholders.

The reconstruction of a major road or interchange in an urban context must be treated as an urban-design project as much as an engineering project. Rue Notre-Dame is not only a service road for the Port of Montreal. Likewise, highways 20, 15 and 720 are not "townless highways," ribbons of asphalt that cut smoothly through the countryside. Where they meet, and before and after that as well, they are parts of the urban landscape, elements of the daily living environment of thousands of Montrealers. Their intersection cannot be designed solely for maximum efficiency of movement and of expenditure; it must be planned as an element of the neighbourhoods around it and not as a barrier between them. The reconstruction of the Turcot Interchange requires the experience of the local resident and the art of the designer as well as the science of the engineer to become a positive contribution to city-building.

The values of speed and efficiency (the former being a subcategory of the latter, really) must be considered together with other values that have inspired planners over the centuries, values such as equity and beauty. How ought the costs and benefits of a plan be distributed? How much money should we spend to make an infrastructure project attractive as well as functional? To what extent should we sacrifice the mobility of long-distance commuters to maintain or improve the mobility of local residents (or vice versa)?

Efficiency itself must be understood broadly. What are the effects that should be maximized? What are the resources that must be utilized sparingly? If allowing for a certain speed of circulation for private automobiles on a highway is efficient in terms of the use of time and in terms of the use of gas, how efficient is that compared to ensuring fast movement for buses or trains that can carry many more passengers at the same time? If creating a new green space by relocating a road helps to make the city greener, how valuable is that when we know that the new trajectory of the road will lengthen travel times and thereby augment the use of scarce resources?

Such questions--in fact all policy and planning questions—cannot be answered by means of numerical computation alone. Answers must be the product of debate in which both social values and scientific facts are being discussed together. Ideally, such a debate must take place in a forum where all stakeholders, from provincial decision-makers to representatives of the neighbourhood population, are present around the same table and are able to speak and be heard (Forester 1989, Healey 1997). All must participate actively in the definition of issues, the identification of values and objectives, and the selection of solutions (Innes 1992, 1995). In such an environment, the professional and the lay person are equal in power.

One brings expert knowledge to the group in support of its collective decision-making; the other contributes an intimate knowledge of the local context and, perhaps, some salutary common sense. All must show respect for perspectives different from their own; all must work together to try and design win-win solutions.

This ideal situation obviously contrasts with traditional arrangements, in which a hierarchy of authority places experts above citizens, provincial planners above municipal colleagues and, to a certain extent, engineers above other professionals. In fact, Québec's Law Respecting Land Use Planning and Development explicitly subordinates local and regional plans to the plans of the province: when a provincial infrastructure project is found to contradict the provision of a regional plan, it is the latter that must be changed, not the former (Québec 2009). Thus, in 2006, the Agglomeration of Montréal had to amend its *schéma d'aménagement* (adopted by the old Montréal Urban Community of 1987) by restoring the extension of Autoroute 25 from Montréal to Laval as a target of the plan, although it had been removed from it in 1988 (MTQ n.d.). In such a context, it is hard to see why provincial planners would voluntarily share power with municipal planners and residents. The law explicitly grants their objectives precedence over the wishes of local actors.

It would take a ministerial resolution to get the MTQ to collaborate with the Ville de Montréal on a major infrastructure project and to participate with local actors in a genuine process of joint decision-making. Such a determination would of course be more likely to be made if the mayor of Montréal flexed whatever political muscle he had to obtain such a degree of collaboration.

On this point, recent experience offers both cause for hope and reason for concern. On the one hand, the city did impose itself as a full partner in the redesign of Rue Notre Dame; on the other hand, it remained far too passive in decision-making on the location of the two new megahospitals. In the first case, it has been engaged in a protracted struggle with the province to make the east-west axis along the Port of Montréal into an urban boulevard rather than a new highway; in the second case, it has simply waited for the province to make its choices known, although the two hospitals are going to have major impacts on their respective environments (McGill University School of Urban Planning 2009).

Leadership by the Minister of Transport and by the Mayor of Montréal is necessary to make the Turcot project into an *urban* project. And that leadership will not be forthcoming, of course, if there is no pressure from the electoral base, if citizens do not make it clear that they expect their elected representatives to act according to the values that they enounce in their speeches.

★ ★ ★

The problem of the Turcot Interchange has no perfect solution, no solution that will satisfy all parties equally. It can please trucking companies better if it limits its response to the claims of local households; it can satisfy environmentalists more fully if it makes them forget that they are also tax-payers. Whatever will be done will represent a compromise among values and a compromise between values and means. How we rank our values and how we allocate our resources in their pursuit is the very stuff of politics. And politics is, or ought to be, a matter of open public debate.

So the problem at hand has no perfect solution. But how we define that problem will be critical to how we solve it and how good a solution we can create (Schön & Rein 1994). The issue can be framed this way: how do we rebuild the intersection of two highways at a cost of less than 1.5 billion dollars? If the two highways are taken as they are and if the budget is taken as a given, then perhaps the best solution is indeed the one proposed by the MTQ, with minor modifications to make public-transit, bicycle and pedestrian movement easier than they are in the current plan.

But the issue can also be framed as follows: how do we reconfigure our transportation system in the southwest of Montréal in the long term? If we can imagine that, one day, highway 720 (Autoroute Ville-Marie) will be taken down east of the Turcot Interchange and highway 15 (Autoroute Bonaventure) will be brought down between Nuns' Island and the Central Business District), then perhaps we need to start planning for a metropolitan core without major highways altogether.

The issue can be placed in a still wider frame: how do we plan the future of the whole southwest portion of Montréal, with its railyards, industrial areas, infrastructure corridors and struggling neighbourhoods? If we see the crumbling of the Turcot Interchange not as a technical challenge but as an opportunity to start a more comprehensive planning exercise for a much larger territory, then perhaps we need to study all possible alternatives as parts of urban-development scenarios. To the preparation of such long-range plans will be brought all the skill of provincial specialists, all the craft of municipal planners and designers and all the talent of Montréalers.

The issue, then, can also be expressed in this manner: how do we organize the process of decision-making in such a way as to bring all this skill, craft and talent together? Formally, that is a question for the Minister and the Mayor to settle. In reality, it is a question for all Montrealers to answer.

### Endnotes

1. Under the heading of "Plans de transport," one reads: "Ils [les plans de transport du Ministère] sont réalisés selon une approche de planification globale. Cette approche tient compte non seulement de tous les modes de transport des personnes et des marchandises, mais aussi des relations des transports avec l'aménagement du territoire, la qualité de l'environnement et le développement socio-économique" (http://www.mtq.gouv.qc.ca/portal/page/portal/ministere/ministere/plans_transport, accessed May 10, 2009).

2. "Le ministère des Transports du Québec a pour mission d'assurer, sur tout le territoire, la mobilité des personnes et des marchandises par des systèmes de transport efficaces et sécuritaires qui contribuent au développement durable du Québec" (http://www.mtq.gouv.qc.ca/portal/page/portal/ministere_en/ministere/organisation, accessed May 10, 2009).

### References

Ellis, Cliff. 1996. "Professional Conflict over Urban Form: The Case of Urban Freeways, 1930 to 1970," in M. C. Sies and C. Silver (eds.), *Planning the Twentieth-Century American City* (Baltimore: The Johns Hopkins University Press), pp. 262-279.

Fogelson, Robert M. 2001. *Downtown: Its Rise and fall, 1880-1950* (New Haven: Yale University Press).

Forester, John. 1989. *Planning in the Face of Power* (Berkeley: University of California Press).

Forester, John, Raphaël Fischler and Deborah Shmueli. 2001. *Israeli Planners and Designers: Profiles of Community Builders* (Albany: State University of New York Press.)

Healey, Patsy. 1997. *Collaborative Planning: Shaping Places in Fragmented Societies* (Houndsmills and London: Macmillan Press).

Innes, Judith E. 1992. "Group Processes and the Social Construction of Growth Management: Florida, Vermont, and New Jersey," *Journal of the American Planning Association* 58(4): 440-453.

Innes, Judith E. 1995. "Planning Theory's Emerging Paradigm: Communicative Action and Interactive Practice," *Journal of Planning Education and Research* **14**(3): 183-9.

Jacobs, Jane. 1961. *The Death and Life of Great American Cities.* (New York: Random House, Vintage Books).

Le Corbusier. 1987 [1925]. *The City of To-morrow and Its Planning* (New York: Dover Publications).

Léonard, Jacques. 1978. "Rencontre avec les maires de la région de Montréal" ("Option préférable d'aménagement pour la région de Montréal") (Québec: Ministère du Conseil exécutif).

Lortie, André (ed.). 2004. *The 60s: Montreal Thinks Big* (Montréal: Canadian Centre for Architecture; Vancouver/Toronto: Douglas & McIntyre).

McGill University School of Urban Planning. 2009. "Making Megaprojects Work for the Community." Retrieved, 11 May 2009, at http://www.mcgill. ca/urbanplanning/mpc.

MTQ. 2009. Web site of the ministry (Québec: Ministère des Transports). Retrieved, 11 May 2009, at http://www.mtq.gouv.qc.ca, accessed on 11 May 1, 2009.

MTQ. n.d. "Le parachèvement de l'autoroute 25 et le schéma d'aménagement en vigueur sur l'île de Montréal" (Québec: Ministère des Transports). Retrieved, 11 May 2009, at http://www.mtq.gouv.qc.ca/portal/page/portal/Librairie/ Publications/fr/centre_affaire/partenariat/a25_consult_mtl.pdf.

Mumford, Lewis. 1964 [1958]. "The Highway and the City," in *The Highway and the City* (New York: Mentor Books, 1964), pp. 244-256.

Québec. 2009. *Loi sur l'aménagement et l'urbanisme. L.R.Q. chapitre A-19.1* (Québec: Gouvernement du Québec). Retrieved, 12 May 2009, at http:// www2.publicationsduquebec.gouv.qc.ca/dynamicSearch/telecharge. php?type=2&file=/A_19_1/A19_1.html.

Schön, Donald A., and Martin Rein. 1994. *Frame Reflection: Toward the Resolution of Intractable Policy Problems* (New York: Basic Books).

Stein, Clarence S. 1957. *Toward New Towns for America* (Cambridge, Mass.: The MIT Press).

Ville de Montréal. 2004a. *Plan d'urbanisme* (Montréal: Ville de Montréal, Service d'urbanisme).

Ville de Montréal. 2004b. Master Plan: Letter from the Mayor (Montréal: Ville de Montréal). Retrieved, 30 April 2009, at http://ville.montreal.qc.ca/portal/ page?_pageid=2762,3099650&_dad=portal&_schema=PORTAL.

Ville de Montréal. 2007. *Réinventer montréal. Plan de transport 2007* (Montréal: Ville de Montréal).

Ville de Montréal. 2009. Plan de transport : les 21 chantiers (Montréal: Ville de Montréal). Retrieved, 12 May 2009, at http://ville.montreal.qc.ca/portal/ page?_pageid=4577,7761620&_dad=portal&_schema=PORTAL.

# Un échangeur dans ma cour : la reconstruction de l'échangeur Turcot et la question de l'intégration urbaine

*Pierre Gauthier*

This chapter discusses the problems associated with the spatial integration of urban highways in densely populated urban areas. A method is introduced that could help to evaluate the impacts of highways on the urban form and the quality of life of the neighbouring populations. It is argued that if highway construction cannot be avoided, such a method could produce criteria to guide our actions. A redevelopment proposal for the Cabot area of Côte-Saint-Paul is used to exemplify the relevance of the approach. This proposal includes an alternative scenario to the proposal by the Québec Ministry of Transportation to rebuild the highway 15 on embankments in the wake of the Turcot interchange reconstruction project.

## Introduction

Ce chapitre traite de la problématique de l'intégration urbaine des infrastructures autoroutières dans les quartiers ouvriers du cœur industriel de Montréal. Ces quartiers centraux densément peuplés, sont ceux-là mêmes qui ont été mis à mal par l'édification des autoroutes urbaines dans les années 1960–70. En plus de faire face à d'importants défis liés à la désindustrialisation, ces derniers sont aujourd'hui confrontés à la perspective de la construction de nouvelles autoroutes qui verrait augmenter fortement l'espace consacré aux transports routiers. C'est ce que préconise par exemple le ministère des Transports du Québec (MTQ) dans son projet de reconstruction de l'échangeur Turcot et de portions significatives des autoroutes 15, 20 et 720 qui y sont rattachées.

Dans les pages qui suivent, j'entends d'abord discuter brièvement en quoi l'initiative du ministère repose sur des prémisses malavisées en ce qu'elle fait largement l'impasse sur les responsabilités environnementales qui nous incombent en cette période de crise des changements climatiques, en choisissant notamment d'ignorer les impacts anticipés d'un tel projet sur le potentiel de requalification et le développement urbain des quartiers traversés. Je traiterai ensuite d'une approche analytique de *lecture du cadre bâti,* qu'il conviendrait d'adopter en vue de produire des connaissances sur le territoire et son évolution. De telles connaissances seraient destinées à informer la conception et l'analyse de scénarios alternatifs de projets de construction d'infrastructures de transport. Je conclurai mon propos en illustrant les retombées concrètes qu'une telle approche permettrait d'anticiper, en présentant une proposition de requalification du secteur Cabot à Côte-Saint-Paul, fondée sur une juste compréhension de la forme urbaine et de l'identité architecturale de ce lieu et bien informée des contraintes relatives à l'intégration d'une autoroute.

## Changements climatiques, aménagement et développement urbain soutenable

Comme en font foi les différentes contributions à cet ouvrage, le projet de réfection de l'échangeur Turcot, rendu public par le MTQ en 2007, est largement critiqué dans ses fondements et prémisses, pour sa conception architecturale, et à l'égard du processus menant à sa réalisation. Plusieurs commentateurs ont soutenu à juste titre, que ce projet d'infrastructure faisait fi des défis posés par la crise des changements climatiques ainsi que des impacts désormais connus et documentés de la pollution atmosphérique imputable au transport routier sur la santé de la population des villes (cf. les chapitres de Bouchard et Ferguson, *et. al.* par exemple).

En matière de lutte contre les changements climatiques et de développement urbain soutenable, il existe au sein des milieux académiques et professionnels dédiés à l'aménagement du territoire, un consensus fort au sujet des principaux moyens à mettre en oeuvre. Ce consensus appelle une réforme en profondeur de nos modes d'occupation et d'utilisation du territoire et un renversement de la tendance en matière de transports urbains. Les propositions touchent trois composantes de l'aménagement : *le mode d'occupation de l'espace* (la distribution spatiale de la population), *l'affectation des sols* (la répartition des usages et des activités urbaines) et *la gestion des transports.*

En réalité, ces trois composantes renvoient à des réalités imbriquées. Ainsi, le lien entre la densité de population et le niveau d'utilisation de l'énergie fossile est abondamment démontré.[1] En clair, une faible densité de population dispersée sur un grand territoire entraîne une plus forte consommation d'énergie à commencer par le carburant associé au transport. De même, s'agissant d'énergie fossile ou autre, des résidences de type pavillonnaire, disséminées sur les larges lots de nos banlieues nord-américaines, sont plus « énergivores » que des habitations plus « ramassées » formant des ensembles plus compacts qui sont moins exposés aux éléments et moins sujets à la déperdition de chaleur par exemple. Dans un même ordre d'idée, il est une qualité première des quartiers densément construits qui mérite attention : ces quartiers peuvent être *marchés* et arpentés à pied. On peut ainsi y avoir accès à courte distance à des services de proximité, tels des commerces sur rue et des services publics ou communautaires, ou à du transport en commun.

Ces constatations, somme toute banales, signifient néanmoins que les gains environnementaux se trouvent en quelque sorte démultipliés dans les quartiers densément peuplés, où, à quantité égale de structures et d'infrastructures architecturales, routières et autres, les personnes, les biens et les services se « trouvent » mutuellement au prix d'un effort considérablement réduit. Les gains se mesurent en termes d'énergie, de temps et de contacts sociaux.

C'est pourquoi des approches telles le *Smart Growth,* le *Pedestrian Oriented Development* (POD), le *Transit Oriented Development* (TOD), ou les *Compact Cities,* de leurs appellations anglo-saxonnes, ont pour préceptes communs : 1. l'augmentation des densités d'occupation de l'espace (densité des populations résidantes et concentration des activités) ; 2. l'intégration plus intime des usages ; 3. le développement des transports actifs (marche, bicyclette) et collectifs. De manière générale, de telles approches visant un développement urbain soutenable visent deux choses en priorité : d'une part, freiner l'étalement urbain et canaliser le développement dans les secteurs déjà urbanisés et pourvus de services (voirie, transport en commun, éducation et services sociaux, commerces, etc.) et, d'autre part, soigner les quartiers centraux, en privilégiant notamment le redéveloppement des friches industrielles qui s'y trouvent. L'idée de base est simple : l'environnement urbain le plus performant du point de vue du développement soutenable s'appuie d'abord sur des quartiers denses, pourvus de services de proximité, et priorise les transports actifs et collectifs comme alternatives à l'automobile individuelle.

Au Québec, quelque 40% des émissions de GES sont imputables aux transports. Les actions les plus vigoureuses sont donc de mise à l'égard de ce secteur si l'on a à cœur d'améliorer notre bilan à ce chapitre. En matière de transport intra-urbain et régional par exemple, le bon sens environnemental et économique nous dicte d'investir nos ressources limitées dans le transport en commun plutôt que dans les autoroutes, partout où faire se peut. Certaines de ces mesures viseront par exemple ceux de ces automobilistes des banlieues qui font seuls la navette, matin et soir, entre ces dernières et le centre-ville et que l'on enjoindra à troquer leur véhicule pour le transport en commun.

Mais il faut savoir que les banlieues, qui sont les héritières lointaines de modèles qui voulaient insuffler de la ville à la campagne et inversement, sont caractérisées, outre leur faible densité, par des réseaux de rues sinueuses que l'on voulait pittoresques et se terminant souvent en culs-de-sac destinés à interdire la circulation de transit. Ces mêmes modèles, du moins dans leur première mouture américaine de banlieues jardins, voyaient la rue comme partie d'un pur réseau technique et fonctionnel ; une rue impropre à soutenir des activités humaines autres que la circulation motorisée, et de ce fait dépourvue de trottoirs. Les humains « non-véhiculés » étaient plutôt destinés à utiliser des réseaux piétonniers verdoyants, indépendants du réseau viaire. Les développeurs immobiliers auront tôt fait d'éliminer ces coûteux appendices piétonniers paysagés. Les piétons banlieusards sont orphelins depuis. Quoiqu'il en soit, le précepte d'urbanisme moderniste qui appelle à une stricte ségrégation spatiale des fonctions urbaines, ne leur donne nulle part où aller à distance de marche. Autre effet pernicieux de ce modèle de développement : les densités prescrites, de même que la nature et la configuration des réseaux de rues de banlieues, ne permettent pas le déploiement d'un service de transport en commun accessible à pied à la majorité de la population à un coût raisonnable. Banlieues et motorisation sont ainsi indissociablement liées depuis les années 1930 en Amérique du Nord. Les banlieusards y sont condamnés aux déplacements motorisés, et le mieux que l'on puisse espérer pour eux aujourd'hui, est de leur permettre de laisser leur véhicule à un point de chute et de transférer vers un moyen de transport collectif de qualité, ce qui leur évitera par exemple les bouchons de circulation du centre-ville.

Ce n'est pas innocemment que j'ai insisté sur le « cas » des banlieues pour évoquer brièvement l'origine de leur forme spatiale. Mon objectif était triple. D'une part, je voulais évoquer les causes sous-jacentes à l'engorgement des centres-villes par les véhicules automobiles. Une telle congestion était inévitable dès lors que le modèle de développement urbain prédominant fut la banlieue jardin, telle qu'on la connaît.

Cette courte démonstration me permet également de tempérer et de mettre en perspective l'idée que les résidents du centre de la ville, qui ont à subir les piètres conditions environnementales associées au flux et reflux quotidien des automobiles, sont victimes d'iniquités sociales. De telles iniquités existent, à n'en point douter, mais il convient tout de même de reconnaître que les banlieusards, qu'ils en soient conscients ou non, sont également les otages d'une situation devenue inextricable. L'automobiliste coincé dans un embouteillage, est non moins exposé à la pollution qu'il contribue à créer. La perspective d'un baril de pétrole de 200$ dans un horizon peut-être pas lointain, peut aider à se représenter les nouveaux coûts individuels et sociaux qui seront, à terme, inévitablement associés à ce mode de développement.

Finalement, mon propos visait à illustrer en quoi le développement du transport en commun au service des banlieues n'est que mesure de temporisation. Quoique essentielle, une telle initiative ne constitue qu'un pis-aller. Ce qu'il convient de faire, afin de casser le cercle vicieux dans lequel on se trouve collectivement enfermés, est d'interrompre derechef la construction de nouvelles banlieues, pour canaliser systématiquement le développement dans les secteurs déjà urbanisés de la ville et au premier titre, dans les quartiers centraux d'avant l'automobile.

Ces derniers, au-delà de la réalité prosaïque des infrastructures, services et autres commodités déjà existantes que l'on voudra y mettre à profit dans un souci d'utilisation raisonnée des ressources, présentent une identité architecturale et des qualités de forme urbaine développées dans la longue durée. Cette identité et cette forme renvoient à un esprit du lieu ou une culture de l'habiter, qui peut servir d'assise à l'élaboration d'une culture urbaine de l'après-automobile. Nouvelle culture urbaine qui se devra sans doute d'être aussi, on commence à le mesurer, une culture de l'après-marchandisation de tout. C'est-là, du moins ce à quoi nous convie un Alberto Magnaghi dans son stimulant ouvrage *Le projet local.*[2]

### Primauté au projet de ville et lecture du cadre bâti

Cette section insiste sur l'importance de fonder toute intervention dans la ville, et a fortiori une intervention aussi perturbatrice que l'introduction d'une autoroute, sur une compréhension approfondie du lieu. La primauté doit toujours être accordée à l'œuvre collective qu'est le projet de ville, qui subsume tous les projets singuliers, quelle que soit leur importance. Cette idée que la ville en projet forme un ensemble dynamique est au fondement d'une méthode, dite de morphogenèse urbaine, développée en Italie. En bref, cette approche nous apprend que l'environnement bâti, comme expression de la culture matérielle, forme une totalité qui évolue dans le temps, selon des règles qu'il nous est possible de comprendre.[3] Le rapport de la société à son territoire se développe dans la longue durée et préside à la production de manières d'habiter, de construire, de partager (au sens de partitionner et d'ordonner) et de se partager socialement l'espace, qui produit des modèles culturels reconnaissables. C'est pourquoi nulle ville n'est à une autre pareille. Dans son expression matérielle, le Montréal ouvrier du 19ième siècle est très différent du Toronto de la même époque, bien que l'urbanisation de ces deux villes ait été soumise à des logiques économiques et industrielles qui sont largement les mêmes.

La physionomie des quartiers ouvriers de Québec et de Montréal doit largement à l'importation en ces lieux de modèles d'habitation rurale d'origine française qui y ont été graduellement transformés pour nous livrer la tradition des « plex », ces logements superposés immédiatement identifiables par les habitants de ces villes. Or, cette architecture est indissociable des tissus urbains dans lesquels elle s'insère.[4] Des rues ont été ouvertes pour accueillir des lots à bâtir, qui ont peu à peu formé des îlots urbains. Le tissu de première édification forme alors une armature dans laquelle l'architecture se déploie spatialement et dans le temps. Si l'architecture peut se transformer rapidement, cela est moins vrai des rues, qui sont beaucoup plus pérennes. Le tissu urbain, à son tour, ne se développe pas dans un vacuum. Sa genèse est contrainte par la trame agricole préexistante ainsi que par la géologie et l'hydrographie du site et, le cas échéant, par de grands ouvrages d'art tels des canaux ou des murs de fortification.

L'urbanisation d'un territoire prend généralement la forme d'un « remplissage » plus ou moins rapide des parcelles agricoles. C'est cette évolution qui vaut à Montréal sa trame de rues orthogonale. Les anciens chemins de campagne, comme les rues Notre-Dame et Saint-Jacques actuelles à Saint-Henri, jouent un rôle déterminant ici ; ils servent de « vecteurs » à l'urbanisation. Le développement s'y déploie d'abord de manière linéaire avant de s'étendre de proche en proche en rase campagne. Du fait de leur destinée particulière, ces chemins sont appelés des parcours-mère du tissu. Ils tirent leur importance, au-delà de leur fonction précédemment décrite, du fait qu'ils connectent le centre de la ville aux nouveaux quartiers, dont ils deviennent souvent le cœur communautaire et commercial. Ces grands parcours de liaison confèrent de l'intelligibilité à la ville pour les usagers, qui en saisissent spontanément l'importance fonctionnelle, si ce n'est symbolique.

On le voit, le territoire humanisé forme un tout imbriqué. L'étude de la ville matérielle, la « lecture » de sa forme physique et spatiale s'opère généralement à quatre niveaux d'échelle spatiale : les édifices, les tissus urbains, la ville et le territoire. L'investigation des éléments qui entrent dans la composition des paysages humanisés ne se limite pas à l'apparence des formes. Elle cherche plutôt à comprendre la logique qui sous-tend leur organisation, à décrire et à expliquer les relations réciproques, difficiles à saisir en raison de leur complexité, qui assurent leur cohérence.[5]

**Figure 2.1a** The Turcot today

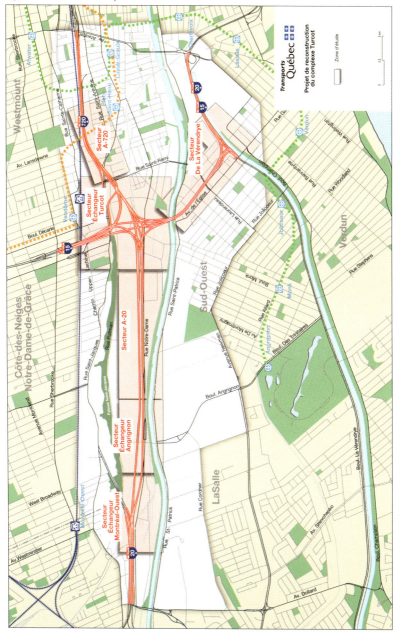

**Figure 2.1b** The MTQ Project

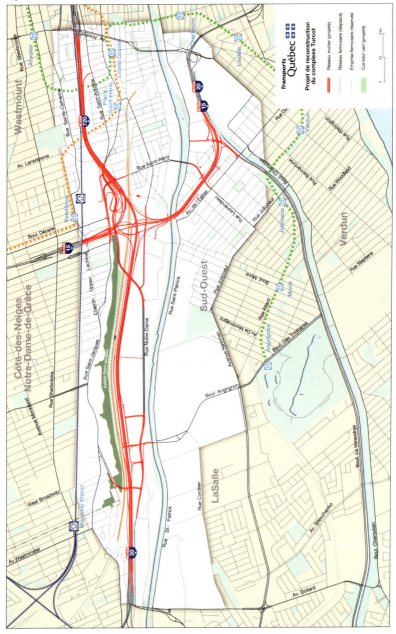

**Figure 2.3** Improve transit between West Island and Downtown Montreal

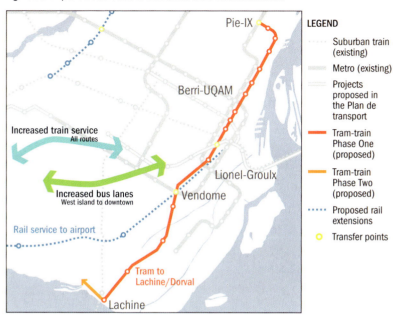

Tyler Baker

**Figure 2.4** MetroExpress

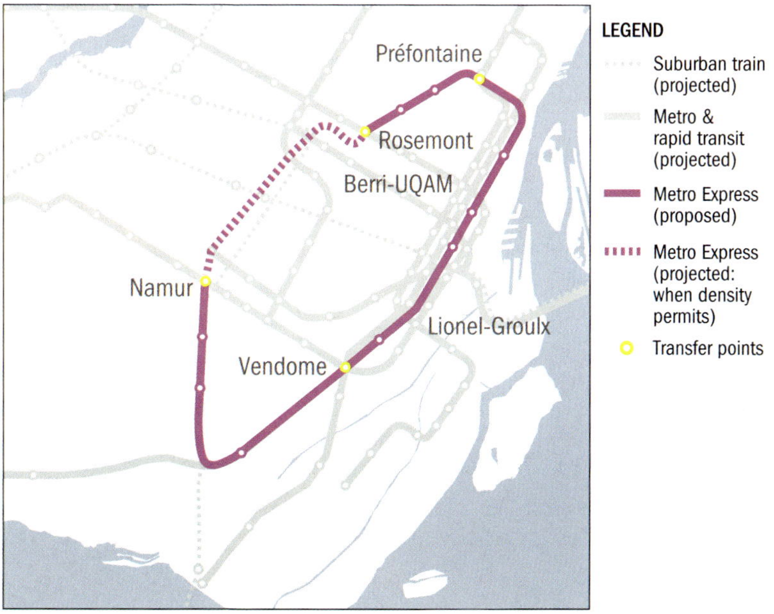

Tyler Baker

**Figure 2.5** Reconnect Westmount to Saint-Henri

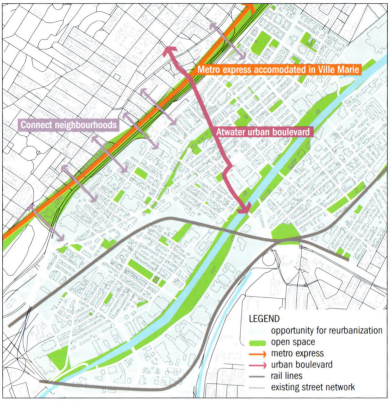

Tyler Baker

**Figure 3.1** Proposed Entrance Pavilion and Belvedere, overlooking the Turcot Interchange and Yards from Terry Fox Park.

Designed by student Anne-Marie Nguyen.

**Figure 3.7** MTQ proposal to move Autoroute 20 and the CN tracks.

MTQ

**Figure 3.8** The new Parc Falaise St. Jacques which includes the Falaise proper enlarged by a green strip (including Rue Pullman) and a bicycle/foot path along the top of the Falaise.

MTQ & Zack Taylor

**Figure 3.9** Section through the Falaise at Terry Fox Park, showing an entrance pavilion straddling the drop in elevation toward the lower level. Design by McGill student Ali Nouri-Nekouri.

**Figure 3.10** Section through Falaise showing tower at left, sliding cable and Pavilion on artificial mount across the St. Pierre river at the bottom of the hill. Design by McGill student Emily Dovbniak.

**Figure 4.1** Map showing potential green paths linking to the Falaise Saint-Jacques

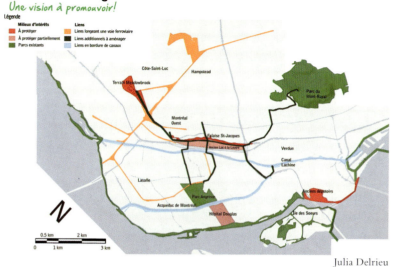

## la Trame verte du grand Sud–Ouest de Montréal
### Une vision à promouvoir!

Légende

**Milieux d'intérêts**
- À protéger
- À protéger partiellement
- Parcs existants

**Liens**
- Liens longeant une voie ferroviaire
- Liens additionnels à aménager
- Liens en bordure de canaux

Côte-Saint-Luc
Hampstead
Terrain Meadowbrook
Parc du Mont-Royal
Montréal Ouest
Falaise St-Jacques
Ancien Lac à la Loutre
Verdun
Canal Lachine
Lasalle
Parc Angrignon
Anciens dépotoirs
Acqueduc de Montréal
Hôpital Douglas
Île des Soeurs

0.5 km   2 km
0   1 km   3 km

Julia Delrieu

**Figure 6.4** Plan d'ensemble du développement projeté au secteur Cabot à Côte-Saint-Paul

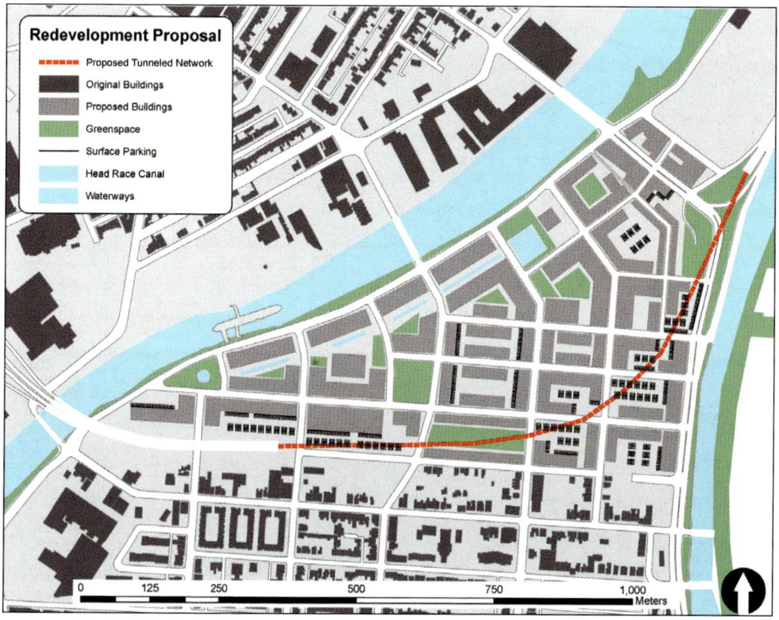

**Redevelopment Proposal**
- Proposed Tunneled Network
- Original Buildings
- Proposed Buildings
- Greenspace
- Surface Parking
- Head Race Canal
- Waterways

0   125   250   500   750   1,000 Meters

M. Budek, E. Goldsmith, R. Sellers, and P. Sobczyk Advanced Urban Laboratory.

**Figure 6.6** Proposition d'enfouissement et de réaménagement des abords de l'autoroute 15 à Côte-Saint-Paul

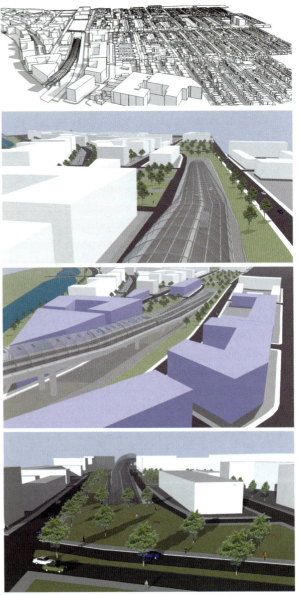

S. Aburihan, M. Cazabon, R. King et D.Stojc, Advanced Urban Laboratory

**Autoroutes et formes urbaines**

Que penser de l'insertion des autoroutes dans le territoire ? La présente discussion ne permet pas de traiter dans le détail des impacts des autoroutes urbaines sur la qualité de la forme urbaine et, par extension, sur la qualité de vie.[6] Certains de ces impacts sont visibles à l'œil nu et s'expriment dans la désolation et la dégradation du cadre bâti que l'on observe aux abords immédiats des autoroutes. Ils tiennent à la fois à la présence matérielle même des ouvrages, par exemple à la zone non aedificandi que constitue généralement leur emprise et aux effets indirects liés à leur aspect visuel généralement rébarbatif et surtout aux conditions environnementales créées par la circulation sur l'autoroute elle-même et sur les quelques rues qui y conduisent. Certains effets sont plus difficiles à saisir empiriquement. Quelles ont pu être par exemple les conséquences sur le cadre bâti et sur la vitalité d'un quartier, de la réquisition d'un ancien parcours-mère ou d'une rue commerciale, pour les transformer en « voies de service » officieuses du réseau autoroutier ? De tels cas s'observent au sujet des rues Notre-Dame et Saint-Antoine dans la Petite-Bourgogne, qui ont sans doute contribué au dépérissement de l'activité commerciale de quartier.

À défaut de pouvoir élaborer comme il se devrait sur les impacts délétères de la construction d'autoroutes au centre de la ville, un sombre constat général s'impose néanmoins : les pratiques d'aménagement de telles infrastructures ont été insensibles à l'histoire et à la spécificité des lieux. En outre, leurs impacts négatifs se sont trouvé accrus, par leur association à une urbanisation de très faible densité, gaspilleuse des ressources et assujettie à une motorisation généralisée, qui étouffe en retour les quartiers centraux plus densément peuplés.

Lorsque l'armature morphologique que constituent les structures spatiales séculaires s'érode, c'est l'historicité et l'identité du lieu qui sont à mettre à la colonne des pertes. Faute de planification adéquate et éclairée ; là où la sédimentation lente conférait aux lieux une intelligibilité et une identité reconnaissables, fruits du croisement culturellement et géographiquement situé entre un territoire et un procès d'urbanisation, on ne trouve trop souvent plus, hélas, qu'un tissu épars, fragmenté et indifférencié.

**Quelles analyses sont de mise, et à quelles fins ?**

L'analyse morphologique de l'intégration des autoroutes urbaines à la ville vise à comprendre le cadre bâti dans lequel ces dernières sont destinées à s'insérer de même que les rapports physiques et spatiaux entre de telles infrastructures et leur environnement d'accueil. Cette analyse a notamment pour objectif de fournir des moyens de minimiser les impacts d'une greffe qui se veut généralement traumatique dans la mesure où les autoroutes, une réalité relativement récente dans l'histoire des villes, ne répondent pas de la même logique de structuration du territoire que les composantes qui renvoient à des modèles culturels et des manières de faire séculaires. Dans cet ordre d'idée, l'analyse morphologique s'avère aussi utile pour évaluer les malformations urbaines qui résultent à court ou moyen terme de la présence d'une autoroute.

Au vue des impacts morphologiques, environnementaux, sociaux et économiques, la question préalable à toute étude sur l'insertion d'une autoroute urbaine devrait toujours être de savoir si cette dernière est nécessaire. Il convient par exemple de se demander si une réduction du nombre de véhicules sur le réseau routier est possible au prix d'une offre accrue de transport en commun. Le cas échéant, l'étude pourra porter sur l'opportunité de substituer à l'autoroute une voie express, tel un boulevard urbain, à l'analyse de l'insertion urbaine de laquelle on pourra alors procéder. Le **Tableau 6.1** par Pierre Larochelle, livre les caractères morphologiques respectifs du boulevard et de l'autoroute.[7] On y remarque qu'une des principales distinctions entre ces deux objets tiens du fait que le boulevard, quoi que soit sa largeur ou sa capacité, est bordé de « bandes de pertinences », ou séries de parcelles portant des bâtiments qui y ont leur adresse civique. C'est cette caractéristique qui fait que contrairement à l'autoroute, le boulevard urbain est assimilable aux modèles culturels associés de longue date à la vie urbaine : en l'occurrence le modèle de la « rue », comme espace de vie et espace social et économique, dont Jane Jacobs nous a montré l'importance cruciale.

S'il s'avère impossible d'écarter la construction (ou la reconstruction) d'une autoroute, l'étude de son intégration dans des secteurs urbanisés devrait comporter des éléments que je livrerai succinctement à des fins d'illustration. Deux types d'analyses sont requis. Le premier aura trait à l'étude de la genèse de l'occupation du territoire touché. Ces analyses auront d'abord pour objet de déterminer les conditions géomorphologiques d'origine (ruisseaux, falaises, etc.), puis celles découlant des premières phases d'occupation sédentaire (les découpages agricoles, les chemins de campagne, les noyaux villageois, la percée d'un canal, etc.). L'étude morphogénétique se portera ensuite sur l'évolution des tissus résidentiels ou spécialisés (les tissus industriels par exemple), depuis la phase d'urbanisation initiale jusqu'à ce jour. Le second type d'analyses portera pour sa part sur les conditions actuelles en vue de dresser un portrait de la distribution

**Tableau 6.1** Caractères comparés : boulevard urbain / autoroute

| | Boulevard urbain | Autoroute |
|---|---|---|
| Nature | Voie intra-urbaine | Voie inter-urbaine |
| Définitions (Selon PRobert) | Boulevard : rue très large, généralement plantée d'arbres | Large route protégée, réservée aux véhicules automobiles, comportant 2 chaussées séparées [...] sans croisements ni passages à niveau |
| Usagers | Automobilistes Cyclistes Piétons | Automobilistes |
| Relation avec le parcellaire et le bâti | AVEC « bandes de pertinence » donc façades principales | SANS « bandes de pertinence » ni adresses civiques |
| Configuration | Avec ou sans terre-plein Avec trottoirs Alignements d'arbres | Chaussées séparées Sans trottoirs |
| Intersections | À NIVEAU relativement rapprochées (180 m max. recommandé) | VIADUCS / ÉCHANGEURS distantes |
| Position relative | Aux confins des tissus urbains (agit comme axe diviseur) | Aux confins des modules territoriaux (barrière infranchissable) |

spatiale et des relations réciproques des différents éléments entrant dans la composition du paysage, tant construit que naturel, dans lequel on envisage d'intervenir. Quelques exemples me permettront d'illustrer le type d'enseignement qu'une telle lecture permet de tirer.

Comme il fut dit précédemment, l'effet conjugué des contraintes topographiques et de celles qui relèvent du système parcellaire issu des premiers découpages agricoles, modèle et conditionne l'occupation du territoire. D'emblée, les limites naturelles (falaise ou cours d'eau par exemple) ou artificielles (tel un canal) ménagent des « aires habitables » ou constructibles. De telles limites forment un réseau de barrières plus ou moins franchissables, qui isolent une région d'une autre. Les points de contacts entre ces aires sont forcément limités. La géométrie des aires habitables se superpose à celles des découpages agricoles et des chemins d'origine. La « charpente » morphologique que produisent ces amalgames informe alors le tracé des grilles de rues, des quadrillages paroissiaux, la localisation des noyaux institutionnels et des services commerciaux de proximité, bref la genèse d'une forme spatiale qui porte la vie communautaire et sociale en général, et qui produit une manière géographiquement et culturellement située d'habiter la ville.

L'ajout d'une barrière infranchissable telle une autoroute altère inévitablement les équilibres spatiaux ainsi que les dynamiques sociales et communautaires qui leur sont associées. L'analyse doit chercher à comprendre ces impacts appréhendés, de même que les impacts physiques et spatiaux des nuisances associées à l'autoroute pour tâcher de les mitiger. Entre autres considérations, on tiendra compte de la position relative de cette dernière dans l'aire morphologique que constitue le quartier. Une position centrale par exemple sera la plus néfaste. Dans le cas où l'autoroute occuperait une position périphérique, on cherchera à mesurer ses effets sur la connectivité des secteurs limitrophes. La même lecture (à l'égard de la position relative et de ses impacts) s'appliquera d'ailleurs aux voies de services et autres voies du réseau viaire affectées par la circulation induite. Une telle analyse s'avère du reste toute aussi pertinente à l'évaluation de l'intégration d'une autoroute, ou d'une voie express à fort débit de type boulevard, dans un secteur en jachère comme la cour Turcot. Dans ce cas, il s'agit de juger de l'impact du positionnement et de la configuration de telles infrastructures sur les secteurs appelés à être développés à l'avenir.

La superposition non contrôlée de barrières entraîne le morcellement des aires « potentiellement » habitables. On y court toujours le risque que les espaces constructibles présentent une superficie qui se situe en deçà des seuils dimensionnels que la pratique raisonnée du design urbain a identifiés comme étant de nature à favoriser un développement socio-communautaire sain et viable.[8] En clair, il faut un espace raisonnable logeant une population de taille suffisante pour justifier des services communautaires et commerciaux de proximité, ou des services de transport en commun par exemple. En deçà de tels seuils, le développement de l'habitation est possible, certes, mais l'automobile devient l'option prédominante pour les déplacements, avec les impacts que l'on sait pour la qualité de l'environnement et de la vie en société. Ces quelques exemples n'épuisent évidemment pas la liste des enjeux relatifs à l'intégration urbaine des autoroutes, mais ils aident à illustrer certains aspects prégnants. La section suivante permettra au lecteur de se représenter un projet de reconstruction d'autoroute doublé d'un effort de requalification urbaine, qui sont fondés sur une approche de lecture morphologique du territoire.

**Figure 6.1** Plan-clé, secteurs Cabot et Galt à Côte-Saint-Paul

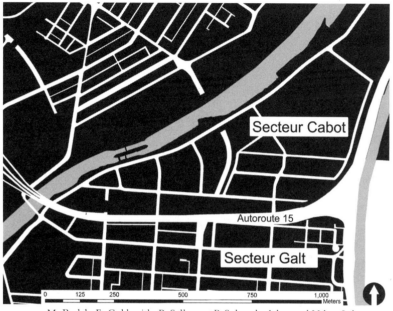

M. Budek, E. Goldsmith, R.Sellers, et P. Sobczyk, Advanced Urban Laboratory

**Le cas de Côte-Saint-Paul**

Cette section illustre le type de retombées concrètes qu'une approche morphologique permet d'anticiper, en présentant une proposition de requalification du secteur Cabot à Côte-Saint-Paul fondée sur de telles assises.
Cette proposition est le fruit des travaux des étudiants en urbanisme du
« Advanced Urban Laboratory » du Département de géographie, urbanisme
et environnement de l'Université Concordia.[9] Elle est marquée du sceau du
développement soutenable et d'une forte volonté de promouvoir l'équité
environnementale et socio-économique.

Le secteur Cabot est une friche industrielle du quartier Côte-Saint-Paul
à Montréal située en bordure du canal Lachine. Cabot jouxte un secteur
essentiellement résidentiel, le secteur Galt (**Figure 6.1**). Les deux ensembles sont séparés par l'emprise de l'autoroute 15, dont la construction dans
les années 1960 a entraîné un appauvrissement de la qualité du cadre bâti
(**Figures 6.2 et 6.3**). L'autoroute ayant atteint la fin de sa durée de vie

**Figure 6.2** Vue du secteur Galt, Côte-Saint-Paul

David Chedore

**Figure 6.3** Vue des secteurs Galt (à l'avant-plan) et Cabot (à l'arrière plan), Côte-Saint-Paul

David Chedore

utile, le ministère des Transports du Québec projette aujourd'hui de faire reconstruire une autoroute sur talus, la solution jugée la plus économique à court et moyen terme.

Des analyses et consultations préliminaires effectuées par les étudiants ont confirmé qu'il est probablement impossible d'éliminer complètement le lien routier de l'autoroute 15 dans ce secteur de la ville. Nos travaux

**Figure 6.5** Proposition d'enfouissement de l'autoroute 15 et de réaménagement des secteurs Galt (à droite) et Cabot (à gauche)

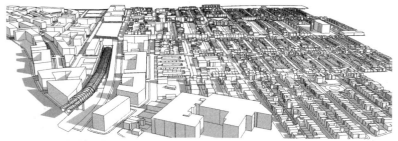

S. Aburihan, M. Cazabon, R. King et D.Stojc, Advanced Urban Laboratory

ont par ailleurs démontré que la reconstruction de l'autoroute sur talus serait néfaste pour la qualité de vie des citoyens du quartier en plus de compromettre gravement le potentiel de développement et de requalification urbaine des secteurs limitrophes, en particulier dans Cabot. Ces mêmes travaux indiquent néanmoins que pour peu que l'on consente un effort réel d'intégration urbaine, rien n'empêche de satisfaire à la fois aux impératifs de la mobilité des biens et des personnes en reconstruisant un axe autoroutier—voire d'y adjoindre un lien ferroviaire—tout favorisant la requalification urbaine d'un milieu de vie et d'un site urbain proprement remarquables.

### Une autoroute mieux intégrée à son environnement

Le projet de reconstruction de l'autoroute sur talus aurait pour effet d'accroître considérablement l'emprise autoroutière en plus d'emmurer le secteur Cabot et d'isoler le secteur Galt plus encore que ce dernier l'est déjà. Les étudiants ont donc développé des scénarios alternatifs qui préconisent, soit l'enfouissement total ou partiel de l'autoroute, soit encore le recouvrement de l'infrastructure actuelle en version recyclée, ou celui d'une autoroute aérienne nouvellement construite. La **Figure 6.4** présente un plan d'ensemble du développement projeté pour le secteur. Une proposition, illustrée par la **Figure 6.5**, témoigne des conditions de redéveloppement que permettrait l'enfouissement de l'autoroute. La **Figure 6.6** (voir la section en couleur) précise le type de développement envisageable aux abords de la portion aérienne et couverte de l'infrastructure à proximité du canal Lachine.

Toutes ces hypothèses auraient pour effet de mitiger les impacts de l'autoroute en permettant d'optimiser le potentiel de développement le long du canal Lachine ainsi qu'aux abords immédiats de l'infrastructure autoroutière elle-même. Elles auraient également pour effet de renforcer les liens entre les secteurs Galt et Cabot, en prolongeant notamment la trame résidentielle du premier vers le second et en retissant le réseau de rues. Ce travail sur la forme urbaine permet d'étendre le tissu résidentiel de Côte-Saint-Paul vers le Nord, et ainsi de lui donner une « façade » sur le canal Lachine et de rendre ce dernier accessible à sa population. Les propositions insistent également sur la nécessité de soigner les abords du canal de l'aqueduc et de tirer profit de cette seconde façade sur canal. Le fait pour un quartier de disposer de deux façades sur canal est une condition rare, qui devrait suffire à elle seule à démontrer l'extraordinaire valeur urbaine, pour ne rien dire de la valeur foncière, des secteurs Galt et Cabot. Cette richesse serait gaspillée de manière irrémédiable par une autoroute sur talus.

Un des objectifs poursuivis a pour but de réduire la fragmentation spatiale du quartier qui résulte de la présence, passée ou actuelle, de barrières physiques que constituent les voies de chemin de fer, les grands ensembles industriels, le tunnel Saint-Rémi et l'autoroute. La fragmentation spatiale a notamment pour conséquences l'enclavement de certains secteurs résidentiels, une distribution fort inégale des services de proximité et de transport en commun, ainsi que des problèmes d'accès piétonnier. Outre la reconstruction souterraine ou aérienne de l'autoroute 15, les divers scénarios incitent à la création d'un réseau d'espaces publics collectifs constitué de rues, places, squares, jardins communautaires et parcs, favorisant les déplacements à pied et encourageant la vie communautaire et la socialisation. Les différentes options étudiées pour la reconfiguration de la trame de rue sont informées du modèle culturel hérité. La construction de nouveaux ponts enjambant le Canal Lachine est également proposée, afin d'assurer une meilleure connexion entre le secteur Cabot, destiné à devenir principalement résidentiel et le quartier Saint-Henri qui lui fait face au-delà du canal.

Les diverses propositions démontrent la possibilité de construire quelque 3000 à 3500 unités d'habitation dans la portion ouest du secteur Cabot. Elles suggèrent la production de formes architecturales variées, destinées à accommoder différents types de ménages. On y adhère à l'objectif d'assurer une diversité de tenures foncières (propriété privée, locatif, logement coopératif et social, logement abordable) qui favorise l'équité socio-économique. Les divers scénarios d'aménagement prévoient la présence : de commerces de quartier sur rue ; de services communautaires (tels une

bibliothèque publique, un centre communautaire, des installations sporti-
ves), ainsi que de secteurs d'emploi (tantôt un secteur de haute technologie,
un marché public, des ateliers d'artistes, un marché d'art et d'artisanat,
un centre de formation de la main-d'œuvre, tantôt encore des secteurs
destinés à l'industrie légère et à l'accueil des industries délocalisées par
le redéveloppement). Le recyclage d'édifices industriels désaffectés à des
fins résidentielles ou commerciales est également au menu, de même que
l'excavation partielle d'un ancien canal d'amenée, destiné ici à animer un
espace public d'envergure.

Un quartier à l'identité unique
Depuis les années 1960, les secteurs Galt et Cabot furent durement
affectés par la désindustrialisation et par les conditions matérielles et environnementales créées par la construction de l'autoroute. Cette évolution
a jeté sur le quartier un voile, qui masque aujourd'hui la richesse de son
patrimoine humain, architectural et urbain. Les travaux conduits par le *Advanced Urban Laboratory* visaient à révéler l'identité de ces lieux. Les propositions d'aménagement s'attardent surtout à la matière concrète de la ville, en
se saisissant des traces du passé industriel (canal d'amenée, anciennes voies
ferrées, bâtiments industriels patrimoniaux) ou en interrogeant la tradition
architecturale et les configurations spatiales résidentielles. Ces formulations
ne servent qu'à suggérer un futur possible : aux citoyens que nous avons
consultés et avec lesquels nous avons collaboré de se saisir maintenant de ce
qu'ils jugeront pertinent dans ce travail. Il nous reste quant à nous à espérer
que notre démonstration amène entre-temps le ministère des Transports
du Québec à reconsidérer son projet destructeur d'autoroute sur talus pour
le secteur.

**Notes**

1. Voir notamment Newman, P., et J.R. Kenworthy. *Sustainability and Cities:
Overcoming Automobile Dependence.* Washington, D.C.: Island Press, 1999.
2. Magnaghi, A. *Le Projet Local.* Traduit de l'italien par M. Raiola et A. Petita,
Architecture et Recherches. Sprimont, Belgique: Pierre Mardaga éditeur, 2003.
3. Voir à ce sujet : Caniggia, G. et G.L. Maffei. *Architectural Composition and
Building Typology. Interpreting Basic Buildings.* Florence : Alinea Editrice, 2001.
4. Sur l'urbanisation industrielle du bassin du Canal Lachine, voir Bliek, D.,
et P. Gauthier. « Understanding the Built Form of Industrialization Along the
Lachine Canal in Montreal ». *Urban History Review—Revue d'Histoire Urbaine*
XXXV, no. 1 (2006): 3-17. Sur l'évolution des formes rurales d'architecture do-
mestique en formes urbaines, voir Gauthier, P. *Le tissu urbain comme forme culturelle:*

*Morphogenèse des faubourgs de Québec, pratiques de l'habiter, pratiques de mise en oeuvre et représentations.* Thèse non-publiée Ph. D., Université McGill, 2003.

5. Pour une description de la méthode d'analyse morphologique appliquée à l'étude des parcours d'accès à la ville, voir Larochelle, P., et P. Gauthier. *Les voies d'accès à la Capitale nationale du Québec et la qualité de la forme urbaine.* Rapport de recherche. Québec: École d'architecture, Faculté d'aménagement, de l'architecture et des arts visuels, Université Laval et Chaire en paysage et environnement, Université de Montréal, 2002.

6. Le lecteur intéressé pourra se référer à P.Larochelle et P. Gauthier op.cit. et à Sénécal, G., J. Archambault, et P.J. Hamel. « L'autoroute urbaine à Montréal: La cicatrice et sa réparation." In *Les espaces dégradés: contraintes et conquêtes*, Sous la direction de G. Sénécal et D. Saint-Laurent, 123-45. Sainte-Foy, Québec: Presses de l'Université du Québec, 2000.

7. Ibid. Larochelle P. et P. Gauthier. 2002 p. 8.

8. Voir à ce propos : Duany, A., E. Plater-Zyberk, et J. Speck. *Suburban Nation: The Rise of Sprawl and the Decline of the American Dream.* New York: North Point Press, 2000.

9. L'auteur désire remercier Kent MacDougall, assistant d'enseignement, Sophie Thiébault de l'Opération Galt et toutes les personnes-ressources de l'arrondissement du Sud-Ouest et des milieux communautaires qui ont contribué aux travaux de l'atelier, ainsi que les étudiants : S. Aburihan, D. Bernardelli, M. Budek, L. Burke, M. Cazabon, D. Chedore, G. Di Cerbo, M. Duchesne, A. H. Durrani, J.-F. Gamble-Beauchamp, J.Gilbert, E. Goldsmith, L. Guglielmino, L. Hang, S. Irani, L. Jia, Y. Katiya, N. Kilmartin, R. King, J. Longo, A. Lucifero, M. Mancini, B. Morell, A. Nelder, R. Sellers, P. Sobczyk, D. Stojc, H. Sugano, X. Y. Wang et M. Yu.

# CHAPTER SEVEN

# Policy Analysis for the Turcot Interchange Reconstruction

*Elham Ghamoushi-Ramandi, Jonathan Moorman, Erika Brown, and M. Munaf Von Rudloff*

Modal shift—an approach that puts mass-transit at the heart of transportation planning—is a sustainability-oriented concept that is gaining world-wide acceptance and acclaim. At the time of writing, a debate is underway in Montreal that has this concept at its heart. The Turcot Interchange, a large inner-city highway junction, is in dire need of repair, replacement or removal. The course of action taken by transportation engineers and decision-makers will reflect either a commitment to sustainability or a step further away from it. If a plan which simply repairs or replaces the structure is implemented, as proposed by the Ministry of Transport of Quebec (MTQ), no modal shift away from auto dependence will occur and the negative effects associated with automobile and truck traffic volumes will continue or increase. If a modal shift-oriented plan is followed, the impacts of traffic in an urban context will decrease.

The purpose of this chapter is to evaluate these two potential courses of action in light of the federal, provincial, and municipal policies that are intended to guide transport and environmental decisions. Three sectors of the environment—noise, transport, and socioeconomic conditions—were used to assess the two alternative plans, the MTQ's repair/reconstruction plan versus a modal shift alternative. Projected impacts were then compared with 124 goals extracted from the relevant policies. The level of compliance between policy goals and environmental impacts was given a quantitative value. Finally, a scoring and weighting system was implemented which produced a numerical value for each alternative course of action. These numerical values represent a "policy score"; in other words, they measure which alternative is most consistent with environmental and transport policy. On a scale of −30 to +30, the Modal Shift alternative received a score of +14.01 (i.e., complies), whereas the current MTQ proposal received a score of −10.92 (i.e., detrimental). Thus, the Modal Shift, or transit-oriented alternative is far more compliant with existing policy goals than the current MTQ proposal.

**Introduction**

In the present international climate of economic crisis, global warming, and the impending end of cheap oil, it is becoming increasingly apparent that our current way of life is not sustainable. Transportation methods, most notably our reliance on the automobile, are often pointed to as major contributors to non-sustainability, especially in urban areas where alternative transport means are readily accessible.

Strategic-level plans and policies in Montreal often contain initiatives to reduce car volumes and increase the use of public modes of transportation. Such policy commitments, and accompanying environmental measures, are apparent at all levels of government, from the municipal (e.g., the City of Montreal's Master Plan and Transport Plan) to the provincial (e.g., the Quebec Ministry of Transport's own Environmental Plan), and federal (e.g., the investment plan in transport) level. Despite this apparent commitment to sustainability, major road and highway projects continue to be approved each year.

An example of this contradiction is the Turcot Interchange reconstruction proposed by the Ministère des Transports du Quebec (MTQ) (MTQ 2008a). This project aims to reinvent the crumbling interchange by lowering its elevated portions to the ground, while simultaneously accommodating an increase in the total number of cars on the structure (from 332,000 vehicles per day to 347,000 vehicles per day, counting all vehicles using all parts of this complex interchange) (SNC-Lavalin/CIMA 2008). The MTQ proposal contained no public transit elements when the research for this chapter was conducted. This chapter sets out a policy assessment of the MTQ's proposal, examining how well it performs against the sustainability objectives and goals contained in current policy documents relevant to the project.

For illustrative purposes, we compare the performance of the MTQ's proposal for the Turcot Exchange with one that could be a viable alternative and explicitly favours more public transit and a 'modal shift' away from automobiles. We chose the "Public Transit" alternative proposed by Pierre Brisset, architect for the Groupe de Recherche Urbaine Hochelaga Maisonneuve (GRUHM), which envisions a reduction in traffic volumes and an eventual removal of the inner-city highway 720, with commensurate increases in public transit (see chapter 2 by Brisset and Moorman).

It is worth reviewing their proposal here. The transit-oriented proposal, described in chapter two, aims to eliminate 68,000 vehicles per day (v.p.d.) on the Ville Marie Expressway by 2016. This objective is achieved by implementing four complementary initiatives: (1) improving public transit service to the West Island; (2) removing ramps and introducing new mass transit links to inner city neighbourhoods to discourage "short" highway trips; (3) introducing drive-alone disincentives such as parking controls and congestion pricing to consolidate these gains; and (4) redesigning the Turcot Interchange to reflect and reinforce lower traffic flows.

Our analysis, described below, shows that the "Public Transit" alternative adheres best to the federal, provincial, and municipal policies that are intended to guide transport and environmental decisions. Our research outlined the regional effects of the MTQ and public transit proposals on transport, noise, and socioeconomic conditions. These effects were then compared with sustainability goals of thirteen relevant policy documents (**Table 7.1**), with the transit-oriented alternative identified as the most strategically appropriate course of action.

## Method

In order to make a comparison of impacts and policy adherence, we developed a series of steps for the assessment of the alternatives. In Table 7.1, we show 'weight' for each policy document, a measure of their importance. First, each document was given a "strategic weight" based on the document's significance to the Turcot Interchange Reconstruction Project, its regional scope, and its relevance to sustainable development goals for the individual sectors (transport, noise and socioeconomic conditions). Some of the documents made no mention of certain environmental sectors, and thus a new "sector weight" was needed that reflected the mention of explicit policy objectives for each sector in the various policy documents (see Table 7.1). For example, Montreal's Master Plan was assigned an overall strategic weight of 14%; however, it has an even higher policy importance relative to other documents with respect to transport (16.87%) and noise (26.42%) since many other documents do not address these sectors.

From the relevant policy documents, 124 goals and objectives for each sector were identified. Policy compliance was then assessed by ranking the performance of each proposal (MTQ and Public Transit) against the policy objective; a simple scale of two pluses (++ or 2) to two negatives (- - or -2) was used to record compliance, with an accompanying rationale provided for each assigned rank (**Table 7.2**). Data on impacts were drawn from existing studies and established methods of environmental impact projections. For example, traffic volumes came from the MTQ (in Atlas des Transports and in the project document submitted by the MTQ for public

**Table 7.1** The 13 policies used in this study and assignment of strategic weights

| Level | Document | Strategic weight | Transport weight | Noise weight | Socioeconomic weight |
|---|---|---|---|---|---|
| Municipal | Montreal Master Plan: November 2004 (Ville de Montréal 2004) | 14% | 16.87% | 26.42% | 15.05% |
| Municipal | Montreal Transport Plan 2008 (Major 2008) | 14% | 16.87% | NA | 15.05% |
| Municipal | Plan de Transport 2007 Mémoire de la Coalition Verte (Green Coalition Verte 2007) | 4% | 4.82% | 7.55% | 4.30% |
| Municipal | Plan d'Action Local pour l'Économie et L'Emploi 2007-2010, RESO (RESO 2007) | 3% | NA | NA | 3.22% |
| Municipal | Transportation Management Plan, Greater Montreal Area (MTQ 2000) | 11% | 13.25% | NA | 11.82% |
| Provincial | The Environmental Policy of MTQ, 1992 (MTQ 1992) | 9% | 10.84% | 16.98% | 9.68% |
| Provincial | Politique sur la bruit routier (MTQ 1998) | 7% | NA | 13.21% | NA |
| Provincial | Quebec and Climate Change, 2008 (MDDEP 2008) | 6% | 7.23% | NA | 6.45% |
| Provincial | Quebec Environmental Quality Act (Gouvernement du Québec 2008) | 7% | NA | 13.21% | 7.52% |
| Provincial | Quebec Public Transit Policy, 2006 (MTQ 2006) | 11% | 13.25% | NA | 11.82% |
| Provincial | Quebec Sustainable Development Strategy 2008-2013 (MDDEP 2007) | 8% | 9.64% | 15.09% | 8.60% |
| Provincial | Transport & Public Health (Drouin 2008) | 4% | 4.82% | 7.55% | 4.30% |
| Federal | Looking to the Future—A Plan for Investing in Canada's Transportation System (The Council of the Federation 2005) | 2% | 2.41% | NA | 2.15% |
| Total | | 100% | 100% | 100% | 100% |

**Table 7.2** Policy compliance of MTQ and Public Transit alternatives: four examples (full table available online: www.gpe.concordia.ca/turcotpolicyanalysis/tables.pdf).

| Document | Goals (examples) | Alternative | | | |
|---|---|---|---|---|---|
| | | MTQ | Rationale: MTQ | Modal Shift | Rationale: Modal Shift |
| Montreal Master Plan: November 2004 (Municipal) | Doubling Montreal's bicycle path network within 7 years (from 400 to 800 km of bike paths) | + | Propose the addition of bicycle paths but no mention of their location or the length of network | – | Does not include any new bicycle initiatives |
| Quebec Sustainable Development Strategy 2008-2013 (Provincial) | To reduce Québec's GHG emissions by 13.8 MT (6% decrease from 1990) | – – | The increase in predicted traffic volumes and automobile use, as well as the lack of public transport initiatives will increase GHG emissions | + | The reduction of the total number of automobiles predicted as well as the increase in public transit plans will decrease GHG emissions |
| Politique sur la bruit routier (Provincial) | Do not exceed a noise level of 55 dBA Leq, 24 h for sensitive areas (residential areas, institutional and recreational) | – – | Based on projected traffic volumes, residential areas neighbouring highway will experience unacceptable noise levels (based on preliminary GIS analysis) | – | Based on projected traffic volumes, some residential areas neighbouring the highway will experience unacceptable noise levels (based on GIS analysis), but there will be a reduction from the current state |
| Montreal Transport Plan 2008 (Municipal) | Reduce dependency on automobile and promote modal shift to public and active transit—reducing costs and commuter times | – | Increase in lanes, increase in capacity | ++ | By third phase, elimination of downtown highway 720 and replaced with public transit options |

**Legend**

| Rank | Score | Description |
|---|---|---|
| ++ | +2 | Exceeds goals |
| + | +1 | Complies |
| = | 0 | Neutral |

| Rank | Score | Description |
|---|---|---|
| – | -1 | Detrimental |
| – – | -2 | Highly Detrimental |
| ? | Removed | Unknown |
| n.m. | Removed | Not Mentioned |

**Table 7.3** Expanded ranking score legend after applying a multiplier of 5

| Sector Score | Total Score | Description |
|---|---|---|
| +10 | +30 | Exceeds goals |
| +5 | +15 | Complies |
| 0 | 0 | Neutral |
| -5 | -15 | Detrimental |
| -10 | -30 | Highly Detrimental |

assessment) (MTQ 2008b; SNC-Lavalin/CIMA 2009), while projected noise and socioeconomic impacts were developed by the authors (www. gpe.concordia.ca/turcotpolicyanalysis/tables.pdf) (Brown et al. 2009).

Based on the ranks assigned to each goal for the two alternative proposals, an overall measure of compliance with the policy document was compiled. The overall score was then multiplied by the specific weight of the document for each sector, providing a weighted score for each document as well as for each alternative. The weighted scores for each document were then totaled in order to determine the total overall score for each sector, for either alternative. For example, MTQ received a total score of -0.92 for the Noise sector, in comparison to the Public Transit proposal's score of +0.39. These values fall within a scale of -2 to +2. In order to expand the range, the total scores were multiplied by 5, which resulted in a new scale of -10 (highly detrimental) to +10 (exceeds goals). The total scores for the Noise sector were therefore transformed into -4.58 for the MTQ alternative and +1.93 for Public Transit alternative.

Lastly, the total score for each alternative was calculated by adding the scores from each sector (a number between -10 and +10), resulting in a final score within the range of -30 to +30. This was done in order to determine the final scores for both the MTQ and Public Transit alternatives, taking into account transport, noise, and socioeconomic sectors. These scores reveal the alternative that complies best with the sustainability policies, plans, and programs that were strategically pertinent to the Turcot Interchange Reconstruction Project (**Table 7.3**).

**Table 7.4** Final result for policy scores for each alternative by sector, as well as total overall score.

| | Alternative | |
| --- | --- | --- |
| Sector | MTQ | Public Transit |
| Transport (Range -10 to +10) | -3.58 | +6.00 |
| Noise (Range -10 to +10) | -4.58 | +1.93 |
| Socioeconomic (Range -10 to +10) | -2.76 | +6.08 |
| Total Score (Range -30 to +30) | -10.92 | +14.01 |

## Results

The policy analysis produced a series of results for the three sectors in each alternative. The final result displayed in **Table 7.4** indicates that the MTQ's alternative falls within the detrimental rank across all sectors with scores of -3.58, -4.58, and -2.76 for transport, noise, and socioeconomic, respectively, within a range of -10 to +10. The Public Transit alternative resulted in values of +6.00, +1.93, and +6.08 for transport, noise and socioeconomic, indicating that this alternative complies better with the policy documents. Within the range of -30 to +30, the MTQ alternative received an overall score of -10.92 and Public Transit proposal received an overall score of +14.01.

**Table 7.5** illustrates the individual scores for each goal and the resulting document scores, using the Montreal Master Plan as an example; all the other policies analyzed can be found online: www.gpe.concordia.ca/turcotpolicyanalysis/tables.pdf.

## Discussion

During the analysis of policy goals, consistent themes emerged. In almost every document, the policymakers aim to enhance quality of life through the reduction of environmental degradation. Public transport, in lieu of automobile use, is routinely advocated. The reduction of noise in urban areas is favoured, and the promotion of socioeconomic benefits through environmentally conscious planning is advanced.

The specific goals we extracted from the policies were also consistent with this theme of improving quality of life through environmental stewardship. These goals were compared with the predicted impacts of each

**Table 7.5** Average scores for the Montreal Master Plan: November 2004 (Municipal). Similar tables produced for all other policy documents can be found online at www. gpe.concordia.ca/turcotpolicyanalysis/tables.pdf

### Transport Policy Analysis

| Document | Goals | MTQ | Public Transit |
|---|---|---|---|
| Montreal Master Plan: November 2004 (Municipal) | 1. Increase the use of public transportation and active modes of transport | -1 | +2 |
| | 2. Reduce the area of off-street parking lots, while applying landscape design and adding green areas | -1 | +1 |
| | 3. Road rebuilding projects should reduce the width of roads and include landscaping and the planting of trees | -2 | +2 |
| | 4. Commitment to reduce greenhouse gases (Kyoto Protocol) | -2 | +1 |
| | 5. Reduce the number of parking spaces, particularly in the Centre | -1 | +2 |
| | 6. Develop new bikeways to serve major activity areas, as well as parking areas for bicycles | +1 | -1 |
| | 7. Establish new public transportation routes in order to facilitate trips between different areas of the City | -1 | +2 |
| | 8. Establish measures to ease traffic flow, such as reducing speed limits, widening sidewalks and designating crosswalks | -1 | +1 |
| | 9. Preserve and enhance the Saint-Jacques escarpment (prevent erosion) | -2 | +2 |
| | 10. The construction of an urban boulevard in the Notre–Dame Street East corridor and the relocation and conversion of the Bonaventure Expressway into an urban boulevard | -2 | +2 |
| | 11. Improve public access to the shorelines, complete the perimeter bikeway and protect the heritage areas and buildings located along the waterside roadway | 0 | 0 |
| | 12. Doubling Montreal's bicycle path network within 7 years (from 400 to 800 km of bike paths) | +1 | -1 |
| | 13. Ensure cycling network is accessible year-round (the white path) | 0 | -1 |

| | | | |
|---|---|---|---|
| 14. Development in partnership of a self-serve bicycle system (operation in 2009 of 2400 bicycles and 300 stations) | | 0 | −1 |

## Noise Policy Analysis

| Document | Goals | MTQ | Public Transit |
|---|---|---|---|
| | Average Score | −0.79 | +0.786 |
| Montreal Master Plan: November 2004 (Municipal) | 1. To limit noise pollution in residential areas | −1 | +1 |

## Socioeconomic Policy Analysis

| Document | Goals | MTQ | Public Transit |
|---|---|---|---|
| Montreal Master Plan: November 2004 (Municipal) | 1. Increase development and density around public transit points such as metro stations | −1 | +2 |
| | 2. Reduce Commuting times through railway/metro expansion and proposed light railway routes | −1 | +2 |
| | 3. Urban development that promotes the use of public transport and active transport—mixed land use | −1 | +1 |
| | 4. Region to Region transport networks to improve commuting time and monthly transport cost | 0 | +2 |
| | 5. Cycling Action Plan—Citywide bikeway | 0 | 0 |
| | 6. Convert disused buildings and empty lots to mixed mode land use especially near transit nodes | +1 | 0 |
| | 7. Reformation of transport assistance and allocate financing | 0 | +2 |
| | 8. Public Transit promotes social equity, thus is targeted to residence, employment and education areas | 0 | +1 |
| | Average Score | −0.25 | +1.25 |

**Table 7.6** Weighted scores for all the 13 policies for all three sectors (see text and Table 1 for explanation of weights).

### Transport Policy Analysis

| Document | MTQ | Public transit | Sector Weight | Weighted Score MTQ | Weighted Score Public transit |
|---|---|---|---|---|---|
| Montreal Master Plan: November 2004 | -0.79 | +0.79 | 16.87% | -0.13 | +0.13 |
| Montreal Transport Plan 2008 | -0.38 | +1.13 | 16.87% | -0.06 | +0.19 |
| Plan de Transport 2007 Mémoire de la Coalition Verte | -0.71 | +0.86 | 4.82% | -0.03 | +0.04 |
| Transportation Management Plan, Greater Montreal Area | -0.57 | +1.43 | 13.25% | -0.08 | +0.19 |
| The Environmental Policy of MTQ, 1992 | -1.00 | +1.50 | 10.84% | -0.11 | +0.16 |
| Quebec and Climate Change, 2008 | -0.67 | +1.00 | 7.23% | -0.05 | +0.07 |
| Quebec Public Transit Policy, 2006 | -1.25 | +1.44 | 13.25% | -0.17 | +0.19 |
| Quebec Sustainable Development Strategy 2008-2013 | -1.50 | +1.50 | 9.64% | -0.14 | +0.14 |
| Transport & Public Health | +1.00 | +1.00 | 4.82% | +0.05 | +0.05 |
| Looking to the Future—A Plan for Investing in Canada's Transportation System | +0.33 | +1.17 | 2.41% | +0.01 | +0.03 |
| Total (Range: -2 to +2) | | | | -0.72 | +1.20 |
| Total (multiplied by 5; Range: -10 to +10) | | | | -3.58 | +6.00 |

### Noise Policy Analysis

| Document | MTQ | Public transit | Sector Weight | Weighted Score MTQ | Weighted Score Public transit |
|---|---|---|---|---|---|
| Montreal Master Plan: November 2004 | -1 | +1 | 26.42% | -0.26 | +0.26 |
| Plan de Transport 2007 Mémoire de la Coalition Verte | +1 | -1 | 7.55% | +0.08 | -0.08 |
| Environmental Policy of MTQ, 1992 | -1 | +1 | 16.98% | -0.17 | +0.17 |
| Politique sur la bruit routier | -0.5 | -0.5 | 13.21% | -0.07 | -0.07 |

| | -2 | -1 | +1 | Sector Weight | Weighted Score MTQ | Weighted Score Public transit |
|---|---|---|---|---|---|---|
| Quebec Environmental Quality Act | -2 | -1 | | 13.21% | -0.26 | -0.13 |
| Quebec Sustainable Development Strategy 2008–2013 | | -1 | +1 | 15.09% | -0.15 | +0.15 |
| Transport & Public Health | | -1 | +1 | 7.55% | -0.08 | +0.08 |
| Total (Range: -2 to +2) | | | | | -0.92 | +0.39 |
| Total (multiplied by 5; Range: -10 to +10) | | | | | -4.58 | +1.93 |

**Socioeconomic Policy Analysis**

| Document | MTQ | Public transit | Sector Weight | Weighted Score MTQ | Weighted Score Public transit |
|---|---|---|---|---|---|
| Plan d'Action Local pour l'Économie et L'Emploi 2007-2010, RESO | -0.67 | +0.50 | 3.22% | -0.02 | +0.02 |
| Plan de Transport 2007 Mémoire de la Coalition Verte | -0.50 | +1.00 | 4.30% | -0.02 | +0.04 |
| Montreal Master Plan: November 2004 | -0.25 | +1.25 | 15.05% | -0.04 | +0.19 |
| Transportation Management Plan: Greater Montreal Area | 0.00 | +1.00 | 11.82% | 0.00 | +0.12 |
| Montreal Transport Plan 2008 | -0.25 | +1.50 | 15.05% | -0.04 | +0.23 |
| Transport & Public Health | -0.50 | +1.00 | 4.30% | -0.02 | +0.04 |
| Environmental Policy of MTQ, 1992 | -1.50 | +2.00 | 9.68% | -0.15 | +0.19 |
| Quebec Public Transit Policy, 2006 | -0.57 | +0.71 | 11.82% | -0.07 | +0.08 |
| Quebec Sustainable Development Strategy 2008-2013 | -0.50 | +1.17 | 8.60% | -0.04 | +0.10 |
| Quebec Environmental Quality Act | -0.50 | +1.00 | 7.52% | -0.04 | +0.08 |
| Quebec and Climate Change, 2008 | -1.50 | +1.33 | 6.45% | -0.10 | +0.09 |
| Looking to the Future: A Plan for Investing in Canada's Transportation System | -1.00 | +2.00 | 2.15% | -0.02 | +0.04 |
| Total (Range: -2 to +2) | | | | -0.55 | +1.22 |
| Total (multiplied by 5; Range: -10 to +10) | | | | -2.76 | +6.08 |

alternative on each environmental sector, and the resulting balance indicates the compliance of each alternative with the policies.

## Policy Analysis—Transport

A clear link between sustainability and transport was made in the documents that were analyzed. The objective to increase the use of public transit and other modes of transport is apparent in nine out of the ten transport documents analyzed. Since the MTQ alternative does not propose any new public transportation initiatives and focuses mainly on automobile transportation, its average score indicated non-compliance. In contrast, the alternative proposal places public transit at the core and proposes many alternate forms of transport: the installation of the "Metro-express;" the Lachine commuter tramway; a railroad shuttle to Dorval; suburban trains, and reserved lanes (Brisset and Moorman 2009). The alternative therefore received an overall positive score (Table 7.6).

The reduction of greenhouse gas emissions is stated as a goal in both the Montreal Master Plan and the Québec Sustainable Development Strategy. The MTQ alternative will clearly not support this objective since the increase in predicted traffic volumes will result in increased emissions, at least in the short term and with existing vehicle technologies. A popular view amongst urban highway proponents is that improved combustion technology will reduce emissions to the point that increased road capacity and higher traffic volumes are environmentally acceptable. However, studies show that increased capacity on highways invariably leads to increased use, congestion, and greater emissions (Noland and Lem 2002). The Public Transit alternative, on the other hand, will reduce the predicted traffic volumes with the removal of the Ville Marie Highway and the implementation of public transit initiatives. This predicted reduction in traffic volumes would reduce $CO_2$ emissions, thus putting this proposal in compliance with emissions goals.

The overall analysis of the Transport sector revealed a score of -3.58 for the MTQ alternative, indicating non-compliance with the 58 goals extracted from the policies. The Public Transit alternative received an overall score of +6.00 for the Transport sector, signifying compliance with the documents analyzed (Table 8.4). These scores are consistent with expectations: the Transport policies all advocate reduced automobile use.

## Policy Analysis—Noise

It can be clearly seen from the comparison of the two alternatives' compliance to noise policy that the MTQ proposal falls, on average, within the detrimental range (negative), while the Public Transit proposal is found to be, on average, compliant (positive) (Table 7.6). While the Public Transit proposal did not comply strongly with noise policies and does not include any noise barriers to mitigate areas experiencing acute noise levels, this alternative will decrease noise levels in the areas surrounding the Turcot Interchange and its adjoining highways, based on preliminary GIS analysis. The MTQ proposal, in contrast, will increase noise levels in the area, but the plan also mentions that noise barriers will be constructed in areas meeting the MTQ's requirements for considering mitigation measures—where noise levels are greater than or equal to 65 dBA, where there are 10 or more houses, and where housing density is at least 30 units per square kilometres (Brown et al. 2009). The proposal does not, however, state where these areas might be. Interestingly, the final results table shows that noise is a significant factor in the MTQ's negative score (-4.58 on a range of -10 to +10). We presume that the MTQ's high negative score is a result of the increased traffic volumes posited by this alternative, notwithstanding the noise barriers.

### Policy Analysis—Socioeconomic Conditions

First, the socioeconomic sector parametres chosen were those likely to be affected by the alternatives for the Turcot. These parametres were: housing prices; commuting times; in- and out-migration; and monthly household travel expenditures. A combination of qualitative methodology, in which general knowledge of generic types of impacts and case studies is used, and quantitative methodology, in which numerical techniques and knowledge of effects which have occurred in similar situations are used, was employed to assess the impacts on the parametres of the Public Transit and MTQ transport initiatives.

We discovered that the Public Transit alternative is far more compliant with socioeconomic policy objectives than the MTQ approach. The four parametres considered are affected by the Public Transit plan in a way that conforms, for the most part, to policy goals (Table 7.6). Housing prices are an exception to this rule: public transit projects typically increase property values in their proximity, while the policy documents specify that housing should remain affordable. Impacts of the Public Transit alternative on commuting times, in-and out-migration, and monthly household travel expenditures are consistent with policy: this alternative received a score within the range of compliance, while the MTQ's alternative fell within the detrimental category. The MTQ's proposal reduces the economic value of inner-city land by sealing it off with embankments, while demolishing approximately 200 units of housing and further reducing the desirability of hundreds of other apartments along its entire route.

Since most of the policies emphasize environmental responsibility, the Public Transit alternative, which advocates reduction of negative effects of transportation, is far more compliant with 58 policy goals. The MTQ alternative, although some of its elements fulfill transport, noise mitigation, and socioeconomic needs, has few environmental measures. The final results of the policy tables confirm this position. This evaluation reflects, quantitatively, that the current MTQ alternative falls far behind other options in enhancing the quality of life of Montreal's residents.

### Improvements to Policy Analysis

A possible criticism of the results presented in this chapter is that different raters may assess the level of compliance of the two alternatives with each goal differently to some degree. For example, the MTQ proposal might be rated "-" instead of "- -" in **Table 7.2** for the goal of reducing Quebec's GHG emissions. Therefore, it would be useful to perform a sensitivity analysis on how much the results change if the alternatives are assessed independently by different raters. However, this apparent variability among raters does not necessarily imply that the assessments are subjective. The 124 goals stated in the 13 policy documents are clear enough that little doubt can exist as to which alternative ranks higher for each particular goal. Overall patterns—and findings—would be the same, even if the quantitative score changes slightly.

A second concern relates to the choice of sectors to examine in depth. Additional sectors, such as Air Quality, Culture/Heritage, Health, and Biology, could be usefully analysed. The analysis of these sectors for each alternative would broaden the scope of impact prediction and assessment.

Likewise, the weighting assigned to each policy document and each sector could be subject to criticism. Again, as an exercise in policy assessment, the broad patterns revealed here would likely hold true—that there are fundamental contradictions between the impacts of highway reconstruction and environmental sustainability aims upheld in numerous local, provincial and federal policies. While the weighting of different policy objectives might shift—and therefore the overall numeric values assigned to each proposal, oversights and shortcomings in the plans (e.g., regarding noise and housing prices in the Public Transit plan; regarding noise, socio-economic impacts and transport objectives in the MTQ plan) revealed through the analysis would still likely be valid.

## Conclusion

An assessment of the compliance of the MTQ's Turcot Interchange Reconstruction project with governmental policy guidelines related to environmental sustainability shows that the proposal does not effectively follow the stated policy goals. The final score for the MTQ alternative was -10.92, which indicates a detrimental overall impact. Reserved bus lanes, bike paths and other measures to promote public and active transit would certainly improve the proposal's compliance with policy, suggesting that the MTQ should seriously undertake a review of its plan in light of existing environmental and transit policies.

Equally important, this chapter shows that an alternative plan that promotes public transit, such as the one put forward by Brisset and Moorman (see chapter 2), outperforms the MTQ's proposals in almost all policy dimensions. The Public Transit alternative received a final score of +14.01 (range -30 to +30), showing compliance with policy objectives. The implications are profound. Development of alternatives for the Turcot Interchange that better match stated local, provincial and federal policies is possible. Rather than commit to the MTQ's plan, with its reliance on highways to maintain current traffic patterns and volumes, this research suggests that investment in the study, elaboration and implementation of policy-compliant alternatives is a more worthwhile endeavour.

## Acknowledgments

This work was part of a student term project in the master's program in Environmental Assessment at Concordia University, Montreal (Brown et al., 2009). We acknowledge our supervisor, Dr. Jochen Jaeger, for his guidance throughout the course of the project. Valuable input was also provided by Pierre Brisset and Jason Prince.

## References

Brisset, P., and J. Moorman. "A Transit-Oriented Vision for the Turcot Interchange: Making Highway Reconstruction Compatible with Sustainability." In *Montreal at the Crossroads: Superhighway, Turcot, and the Environment*, edited by Pierre Gauthier, Jochen Jaeger and Jason Prince. Montreal, New York, London: Black Rose Books, 2009.

Brown, Erika, Elham Ghamoushi-Ramandi, Jonathan Moorman, and Munaf von Rudloff. "Strategic Environmental Assessment of the Turcot Interchange Reconstruction." Montreal, Quebec: Concordia University, 2009, unpublished report.

Drouin, Louis. "Transport Et Santé Publique Enjeux Et Solutions." Montreal, Quebec: Presented at the Colloque Écosanté, ACFAS, May 6, 2008.

Gouvernement du Québec. "Quebec Environmental Quality Act." Quebec City, Quebec: Gouvernement du Québec, 2008.

Green Coalition Verte. "Plan De Transport 2007; Mémoire De La Coalition Verte." Montreal, Quebec: Green Coalition Verte, 2007.

Major, Francois. «Montreal Transport Plan: A Strategic Approach to Sustainable Transportation.» Montreal, Quebec: Ville de Montréal, 2008.

(MDDEP) Ministère du Développement durable, de l'Environnement et des Parcs. "A Collective Commitment: Government Sustainable Development Strategy, 2008-2013." Quebec City, Quebec: Ministère du Développement durable, de l'Environnement et des Parcs, 2007.

(MDDEP) Ministère du Développement durable, de l'Environnement et des Parcs. "Québec and Climate Change: 2006-2012 Climate Change Action Plan." Quebec City, Quebec: Ministère du Développement durable, de l'Environnement et des Parcs, 2008.

(MTQ) Ministère des Transports du Quebec. "Atlas Des Transports." (2008a), http://transports.atlas.gouv.qc.ca/Infrastructures/InfrastructuresRoutier.asp.

(MTQ) Ministère des Transports du Quebec. "Environmental Policy of the Ministère Des Transport Du Québecenvironmental Policy of the Ministère Des Transport Du Québec." Quebec City, Quebec: Ministère des Transports du Quebec, 1992.

(MTQ) Ministère des Transports du Quebec. "Politique Sur La Bruit Routier." Quebec City, Quebec: Ministère des Transports du Quebec, 1998.

(MTQ) Ministère des Transports du Quebec. "Québec's Public Transit Policy." Quebec City, Quebec: Ministère des Transports du Quebec, 2006.

(MTQ) Ministère des Transports du Quebec. « Reconstruction of the Turcot Complex, in Montréal. » (2008b), http://www.mtq.gouv.qc.ca/portal/page/portal/entreprises_en/zone_fournisseurs/c_affaires/pr_routiers/reconstruction_complexe_turcot_mtl.

(MTQ) Ministère des Transports du Quebec. "Transportation Management Plan, Greater Montréal Area: Priority Intervention Strategy." Quebec City, Quebec: Ministère des Transports du Quebec, 2000.

Noland, Robert B., and Lewison L. Lem. «A Review of the Evidence for Induced Travel and Changes in Transportation and Environmental Policy in the Us and the Uk.» *Transportation Research Part D: Transport and Environment* 7, no. 1 (2002): 1-26.

(RESO) Regroupment Économique et Social du Sud-Ouest. « Plan D'action Local Pour L'économie Et L'emploi 2007—2010. » Montreal, Quebec: Regroupment Économique et Social du Sud-Ouest, 2007.

SNC-Lavalin/CIMA, Consortium. « Projet De Reconstruction Du Complexe Turcot; Impacts Sonores Rapport Sectoriel, Annexe H: Données De Circula-

tion (Djme) Prévues En 2016 Avec Le Projet De Reconstruction Du Complexe Turcot.» Ministère des Transports du Quebec, October 29, 2008.
The Council of the Federation. «Looking to the Future—a Plan for Investing in Canada's Transportation System.» Ottawa, Ontario: The Council of the Federation, 2005.
Ville de Montréal. "Montréal Master Plan." Montreal, Quebec: Ville de Montréal, 2004.

## Appendix

**Table 7.7** shows all 124 goals extracted from the 13 policy documents and the average scores for the two alternatives, for the three sectors (58 goals for Transport, 8 goals for Noise, 58 goals for Socio-economic). **Table 7.8** showing the full list of all rationales used in assessing the two alternatives in this study is available at the following website: www.gpe.concordia.ca/turcotpolicyanalysis/tables.pdf."

**Table 7.7** Average scores for all 13 documents for the sectors Transport, Noise, and Socio-economic.

### Transport Policy Analysis

| Document | Goals | MTQ | Modal Shift |
|---|---|---|---|
| Montreal Master Plan: November 2004 (Municipal) | 1. Increase the use of public transportation and active modes of transport | -1 | +2 |
| | 2. Reduce the area of off-street parking lots, while applying landscape design and adding green areas | -1 | +1 |
| | 3. Road rebuilding projects should reduce the width of roads and include landscaping and the planting of trees | -2 | +2 |
| | 4. Commitment to reduce greenhouse gases (Kyoto Protocol) | -2 | +1 |
| | 5. Reduce the number of parking spaces, particularly in the Centre | -1 | +2 |
| | 6. Develop new bikeways to serve major activity areas, as well as parking areas for bicycles | +1 | -1 |
| | 7. Establish new public transportation routes in order to facilitate trips between different areas of the City | -1 | +2 |
| | 8. Establish measures to ease traffic flow, such as reducing speed limits, widening sidewalks and designating crosswalks | -1 | +1 |
| | 9. Preserve and enhance the Saint-Jacques escarpment (prevent erosion) | -2 | +2 |
| | 10. The construction of an urban boulevard in the Notre-Dame Street East corridor and the relocation and conversion of the Bonaventure Expressway into an urban boulevard | -2 | +2 |
| | 11. Improve public access to the shorelines, complete the perimeter bikeway and protect the heritage areas and buildings located along the waterside roadway | 0 | 0 |
| | 12. Doubling Montreal's bicycle path network within 7 years (from 400 to 800 km of bike paths) | +1 | -1 |
| | 13. Ensure cycling network is accessible year-round (the white path) | 0 | -1 |
| | 14. Development in partnership of a self-serve bicycle system (operation in 2009 of 2400 bicycles and 300 stations) | 0 | -1 |
| | Average Score | -0.79 | +0.786 |

| Source | Policy | | |
|---|---|---|---|
| Montreal Transport Plan, 2008 (Municipal) | 1. Improve the STM's services to increase usage by 8 % in 5 years | -1 | +2 |
| | 2. Build a tramway network in the core of the agglomeration | -2 | +2 |
| | 3. Initiatives to have a Train shuttle to airport | -2 | +2 |
| | 4. Maintain and complete the road network | +2 | +2 |
| | 5. Priority measures for 240 kilometres of arterial roads | +2 | -2 |
| | 6. Commuter railroad in the East | 0 | 0 |
| | 7. Rapid transit system with exclusive right-of-way lanes | -1 | +2 |
| | 8. Modernize and expand Métro system eastward | -1 | +1 |
| | *Average Score* | *-0.38* | *+1.125* |
| Plan de Transport 2007 Mémoire de la Coalition Verte (Municipal) | 1. Encourage use of subways, trains and light rail vehicles | -2 | +2 |
| | 2. Moratorium on new road/highway construction until the current network is fixed | +1 | -1 |
| | 3. Focus on rail transit versus buses on trunk lines | -1 | +1 |
| | 4. Conservation of strategic rail corridors for future transit use | +1 | +1 |
| | 5. Designation of the CPR line, or something paralleling that line as the route for the airport shuttle | -2 | +1 |
| Plan de Transport 2007 Mémoire de la Coalition Verte (Municipal) | 6. Electrify existing diesel operated commuter rail lines and increase their frequency | -1 | +1 |
| | 7. Encourage more freight on the railways | -1 | +1 |
| | *Average Score* | *-0.71* | *+0.857* |

| | | | |
|---|---|---|---|
| Transportation Management Plan, Greater Montreal Area (Municipal) | 1. Emphasize initiatives that foster the revitalization and consolidation of the urban core | -1 | +2 |
| | 2. Promote mass transit & transportation management to limit the drawbacks of the automobile (eg. noise) | -1 | +2 |
| | 3. Give priority to reinforcing and modernizing existing transportation networks | +1 | +2 |
| | 4. Promote broader use of mass transit: new metro lines, commuter trains, reserved lanes, park and ride | -2 | +2 |
| | 5. Manage transport demand instead of reacting to it | -2 | +2 |
| | 6. An integrated service strategy for the east end of the Greater Montreal Area | -1 | +1 |
| | 7. Priority initiatives pertaining to the road network in and leading to urban core | +2 | -1 |
| | *Average Score* | *-0.57* | *+1.429* |
| The Environmental Policy of Quebec's MTQ, 1992 (Provincial) | 1. Improve the complementarity of the various modes of transport | -1 | +2 |
| | 2. Design transportation infrastructures with a view to favouring the development of the living environment | -1 | +1 |
| | *Average Score* | *-1* | *+1.5* |
| Quebec and Climate Change, 2008 (Provincial) | 1. Encourage the development and use of transportation alternatives | -1 | +2 |
| | 2. Encourage the development and use of public transit | -2 | +2 |
| | 3. To develop networks of safe bicycle lanes that run from residential neighbourhoods to employment centres like downtown areas, industrial parks, shopping centres, etc. | +1 | -1 |
| | *Average Score* | *-0.67* | *+1* |

| | | | |
|---|---|---|---|
| Quebec Public Transit Policy, 2006 (Provincial) | 1. A 8% increase in public transit ridership by 2012 | -1 | +2 |
| | 2. Improve quantity/quality of public transit services | -1 | +2 |
| | 3. Backing alternatives to automobiles | -1 | +2 |
| | 4. Improving energy efficiency of passenger transportation by road | -2 | +1 |
| | 5. Ensuring the environment, society, and economy are factored into every transport decision | -2 | +1 |
| | 6. A 10% reduction in present-day energy consumption by 2015 | -2 | +1 |
| | 7. A 16% increase in the supply of public transit services | -2 | +2 |
| | 8. To make cycling more compatible with another mode of transportation | +1 | +1 |
| | *Average Score* | *-1.25* | *+1.444* |
| Quebec Sustainable Development Strategy 2008-2013 (Provincial) | 1. To reduce Québec's GHG emissions by 13.8 MT (6% decrease from 1990) | -2 | +1 |
| | 2. Support Quebec Public Transit Policy | -1 | +2 |
| | *Average Score* | *-1.5* | *+1.5* |
| Transport & Public Health (Provincial) | 1. Safer Roads | +1 | +1 |
| | *Average Score* | *+1* | *+1* |
| Looking to the Future—A Plan for Investing in Canada's Transportation System (Federal) | 1. Encompassing all modes of transportation in a balanced and integrated way | -1 | +2 |
| | 2. Improving safety, security and efficiency on corridors serving strategic gateways and key economic nodes | +1 | +1 |
| | 3. Recognizing the strategic role of urban centres & urban transit and eliminating bottlenecks within & between cities | +2 | +1 |
| | 4. Facilitating interprovincial/territorial, international...tourist traffic | 0 | +1 |

| Document | Goals | MTQ | Modal Shift |
|---|---|---|---|
| Looking to the Future—A Plan for Investing in Canada's Transportation System (Federal) | 5. Improving access to strategic transportation components that currently have aging, congested or absent highway connections | +1 | +1 |
| | 6. Promoting innovation and efficiency in transport | -1 | +1 |
| | *Average Score* | *+0.33* | *+1.167* |

### Noise Policy Analysis

| Document | Goals | MTQ | Modal Shift |
|---|---|---|---|
| Montreal Master Plan: November 2004 (Municipal) | 1. To limit noise pollution in residential areas | -1 | +1 |
| Plan de Transport 2007 Mémoire de la Coalition Verte (Municipal) | 1. Improvements like: Noise/sound barriers | +1 | -1 |
| Environmental Policy of Quebec's MTQ. 1992 (Provincial) | 1. Reduce noise and other forms of pollution generated by the construction, use and maintenance of transportation infrastructures | -1 | +1 |
| Politique sur la bruit routier (Provincial) | 1. Do not exceed a noise level of 55 dBA Leq. 24 h for sensitive areas (residential areas, institutional and recreational) | -2 | -1 |
| | 2. Construct barriers in areas where noise levels are 65 dBA Leq 24 h and include at least 10 housing units with a density of 30 dwelling units per kilometer | +1 | 0 |
| | *Average Score* | *-0.5* | *-0.5* |

| Document | Goals | MTQ | Modal Shift |
|---|---|---|---|
| Quebec Environmental Quality Act (Provincial) | 1. Prescribed standards for noise intensity | -2 | -1 |
| Quebec Sustainable Development Strategy 2008-2013 (Provincial) | 1. Increase standard of living (eg. limit noise exposure) | -1 | +1 |
| Transport & Public Health (Provincial) | 1. Decrease Noise | -1 | +1 |

**Socioeconomic Policy Analysis**

| Document | Goals | MTQ | Modal Shift |
|---|---|---|---|
| Plan d'Action Local pour l'Économie et L'Emploi 2007-2010, RESO (Municipal) | 1. Creation of a healthy socio-economic environment | +1+ | +1 |
| | 2. Creation of jobs and local business opportunities | +1 | 0 |
| | 3. Raise the cultural value of the area | -1 | +1 |
| | 4. Commitment to sustainable development in social, economic, and environmental realms | -2 | +1 |
| | 5. Develop affordable housing simultaneously with condominium options to meet the needs of a rapidly expanding population | 0 | 0 |
| | 6. Improve public transit to neglected zones in collaboration with residences and local businesses | -1 | +2 |
| | 7. Enlarge social housing options in response to different needs | 0 | 0 |
| | 8. Diversify housing options under the perspective of mixed social conditions, without penalizing social and community housing | -1 | 0 |
| | 9. Limit rent increases in the area | -1 | -1 |

| Plan | | -1/0 col | +1/+2 col |
|---|---|---|---|
| Plan d'Action Local pour l'Économie et l'Emploi 2007-2010, R.ESO (Municipal) | 10. Improve and develop services in close proximity to residential areas | -1 | +1 |
| | 11. Prioritize public transit development in order to unblock traffic congestion in the industrial sector of Point St Charles | -2 | +2 |
| | 12. Maintain a minimum of 30% of "affordable" housing, with half of that being social housing, in every residential project of over 200 units | -1 | -1 |
| | *Average Score* | *-0.67* | *+0.50* |
| Plan de Transport 2007 Mémoire de la Coalition Verte (Municipal) | 1. Reduce commuting times by means of augmenting commuter rail links | -1 | +2 |
| | 2. Implementation of regulations for railway lines near housing (barriers etc) to mitigate property value decline | 0 | 0 |
| | *Average Score* | *-0.50* | *+1.00* |
| Montreal Master Plan: November 2004 (Municipal) | 1. Increase development and density around public transit points such as metro stations | -1 | +2 |
| | 2. Reduce Commuting times through railway/metro expansion and proposed light railway routes | -1 | +2 |
| | 3. Urban development that promotes the use of public transport and active transport—mixed land use | -1 | +1 |
| | 4. Region to Region transport networks to improve commuting time and monthly transport cost | 0 | +2 |
| | 5. Cycling Action Plan—Citywide bikeway | 0 | 0 |
| | 6. Convert disused buildings and empty lots to mixed mode land use especially near transit nodes | +1 | 0 |
| | 7. Reformation of transport assistance and allocate financing | 0 | +2 |
| | 8. Public Transit promotes social equity, thus is targeted to residence, employment and education areas | 0 | +1 |
| | *Average Score* | *-0.25* | *+1.25* |

| | | | |
|---|---|---|---|
| Montreal Transport Plan 2008 (Municipal) | 1. Reduce dependency on automobile and promote modal shift to public and active transit—reducing costs and commuter times | -1 | +2 |
| | 2. Manage parking at a strategic level to mitigate traffic and automobile dependency in dense areas | 0 | +1 |
| | 3. Construction of pedestrian oriented areas therefore promoting intensification | 0 | +1 |
| | 4. Allocate Urban transit funding and reshape transit assistance | 0 | +2 |
| | *Average Score* | *-0.25* | *+1.50* |
| Transportation Management Plan: Greater Montreal Area (Municipal) | 1. Tax breaks, subsidies for mass transit use: Both to individuals and employers with transit benefit programs | 0 | +1 |
| | *Average Score* | *0.00* | *+1.00* |
| Transport & Public Health (Provincial) | 1. Provision of solutions to ameliorate health and economic standing with regards to transport in budget alignment | -1 | 0 |
| | 2. Reduction in speed and congestion and a shift to public transit alternatives and active transit | -1 | +1 |
| | 3. Improve alternatives to all members of society for transportation, allocate funding | 0 | +2 |
| | 4. Reduce parking access, thus looking to reduce urban traffic, replaced with cost effective transit | 0 | +1 |
| | *Average Score* | *-0.50* | *+1.00* |
| Environmental Policy of Quebec's MTQ, 1992 (Provincial) | 1. Integrate land-use and transport in a way which encourages modal shift | -2 | +2 |
| | 2. Accommodate people's growing need for rapid transport through systems which do not involve private automobile use | -2 | +2 |
| | 3. Combat urban sprawl through better land-use planning and transport integration | -1 | +2 |
| | 4. Reverse the trends of populations moving towards linear infrastructures and away from downtown cores | -1 | +2 |
| | *Average Score* | *-1.50* | *2.00* |

| | | | |
|---|---|---|---|
| Quebec's Public Transit Policy, 2006 (Provincial) | 1. Promote the role of public transit as a lever for economic development | 0 | +1 |
| | 2. Encourage businesses to locate favourably for public transit access, and to provide workers with incentives to use public transit | -2 | +2 |
| | 3. Reduce the monetary cost of congestion by reducing commuting times | -1 | +2 |
| | 4. Ensure fair distribution of the burden of travel costs through taxation of gasoline and fees paid from vehicle licenses | 0 | 0 |
| | 5. Allocate greater government funding to remote and sparsely populated regions in order to give residents wider transport choices | 0 | 0 |
| | 6. Keep public transit fares within a range which does not inhibit use or restrict access to the service | 0 | -1 |
| | 7. Sustain the economy while improving the quality of urban life and the mobility of low-income earners | -1 | +1 |
| | *Average Score* | *-0.57* | *+0.71* |
| Quebec Sustainable Development Strategy 2008–2013, (Provincial) | 1. To link environmental protection, social progress, and economic efficiency together under the auspices of sustainable development | -1 | +1 |
| | 2. To encourage the economies of Quebec and its regions to effectively commit to innovation, prosperity, social progress and respect for the environment | 0 | +1 |
| | 3. "Polluter pays:" the costs of measures to prevent, control, and mitigate environmental damage are to be paid by the polluter (i.e. drivers, industrialists) | -1 | +1 |
| | 4. Adjust to demographic changes by fostering economic prosperity through innovation | +1 | +1 |
| | 5. Practice integrated, sustainable land-use and development | -2 | +2 |
| | 6. Ensure the welfare of families and the promotion of conditions favourable to family life | 0 | +1 |
| | *Average Score* | *-0.50* | *+1.17* |

| | | | |
|---|---|---|---|
| Quebec Environmental Quality Act (Provincial) | 1. Maintain citizens' right to a healthy environment and the protection of all living species inhabiting it. | -1 | +2 |
| | 2. Regulate for the protection of integrity of the land and its use; rehabilitate where necessary to maintain this integrity | 0 | 0 |
| | *Average Score* | *-0.50* | *+1.00* |
| Quebec and Climate Change, 2008 (Provincial) | 1. Deal with climate change through two main strategies: avoid emissions and adapt to changing climatic conditions | -1 | +1 |
| | 2. Maximize Quebec's economic competitiveness through optimizing energy efficiency | -1 | +1 |
| | 3. Implement a greenhouse gas emission cap and trade system for certain economic sectors which are heavy GHG emitters | -1 | +1 |
| | 4. Reduce emissions from Quebec's transport sector, which in 2005 accounted for 39% of total provincial greenhouse gas emissions | -2 | +2 |
| | 5. Encourage the development and use of public transit and reduce the daily use of single passenger vehicles | -2 | +2 |
| | 6. Encourage the implementation of multi-modal projects for the transportation of merchandise, reducing the role of trucks in goods transport and promoting the use of rail | -2 | +1 |
| | *Average Score* | *-1.50* | *+1.33* |
| Looking to the Future: A Plan for Investing in Canada's Transportation System (Federal) | 1. To recognize and develop the critical link between transportation infrastructure and the economy | -1 | +2 |
| | 2. To pursue opportunities to promote awareness of the importance of sustainable urban transport and transportation choices to the economic and social lives of Canadians | -1 | +2 |
| | *Average Score* | *-1.00* | *+2.00* |

# CHAPTER EIGHT

# The Health Effects of Road Traffic— A Brief Overview

*N. Meaghan Ferguson, Robert J. Moriarity, Frederic Gagnon and Melanie J. McCavour*

Scientific research conducted over the last 30 years has accumulated evidence for the numerous impacts of road traffic on health. These include increased risk of heart attack, increased blood pressure, asthma, wheezing in infants, sleep deprivation, reduced birth weights, and psychological effects such as stress, anxiety, and diminished memory capacity. These health effects particularly impact people of lower socio-economic status because they are more likely to be exposed to the worst air quality, the highest noise levels, and overall poor environmental quality than those with higher socio-economic status. The effects are more pronounced with increasing proximity to highways, where a 200-metre buffer from the highway has been identified as a zone of elevated health risks. This has important implications for the environmental assessment of road projects. For example, air quality in the vicinity of Montreal's Turcot Interchange cannot be predicted on the basis of air quality stations located several kilometres away, as was attempted in the Environmental Assessment report presented by the Ministère des Transports du Québec.

## Introduction

This chapter provides an overview of the impacts of road traffic on the health of residents living within a 200-metre distance from a major roadway based on relevant scientific literature. The distance of approximately 200-metres has been identified as the width of the buffer where negative health effects due to air pollution and noise have clearly been demonstrated. The emphasis of this chapter is on the effects of noise and air pollution on quality of life as well as the link with socioeconomic status.

## Noise and Vibrations

Since the late 1970s, research has shown a strong influence of noise, the invisible pollutant, on physical and mental health. Detailed investigations have revealed a strong connection between noise and cardiovascular, mental, and other health effects.

Road noise includes any noise generated by vehicles moving on a road surface. Noise could emanate from the vehicles themselves through engine noise or braking, and from the contact of the tires with the road surface. The main factors to consider are the volume and speed of traffic, and the number of trucks in the traffic flow; trucks generally create more noise than the typical car. Road noise levels also depend on how close the perceiver is to the road (Washington State Department of Transportation 2009).

Exposure to road noise at an average distance of 200 metres or less poses a risk to human health (Genereux et al. 2007). Any amount of noise that exceeds 50 decibels (dBA) for more than 24 hours is enough to cause stress on the physical and mental well being of those living in affected areas (Bluhm 2004). Furthermore, with the high cost of property involved in highways and road construction, it is not always feasible to construct buffer zones around roads; therefore, residential areas often end up in this 200-metre area. Certainly, this is the case with Montreal's *Village des Tanneries*, a small neighbourhood immediately abutting the Turcot Interchange: it was there long before this structure and the Ville Marie expressway (Autoroute 720) were constructed.

Of those individuals living near a road or highway that has a continuous noise level over 50 dBA, with periodic maximums reaching 70 dBA, there is an approximately 1.5 times increase in the risk of a heart attack, and for those individuals living there more than 10 years there is a 2-fold risk increase (Babisch et al. 2005). The risk of hypertension increases for those individuals living in the vicinity of a busy highway by 1.4 times per 5 dBA increase, with a maximum risk increase of approximately 2.5 times (Bluhm et al. 2007). The literature supports the causal link between road noise and cardiovascular problems; therefore, it is safe to conclude that chronic exposure to high levels of traffic noise increases the risk of physical health problems.

The continuous noise of a highway clearly has a profound effect on physical health; however its effects on mental health are also widely noted (Lercher 1996; Muller-Wenk 2002). Experts agree that anyone living within 200 metres from a busy road or highway may have their mental health jeopardized after residing in the area for a period of more than one year. Impacts on health from noise-induced mental stress can result in memory loss, that is especially noticeable in children and long-term residents, an increased level of anxiety and depression, and in some cases elevated levels of substance abuse (Passcher 2000). Studies have also noted that those affected may not notice continuous noise after a certain amount of time, however the effects on mental health are present (Stansfield 2000).

The implications from non-auditory noise (i.e., vibrations) on the health of those living near a busy road or highway are still under investigation; however, studies show a general trend of changes in hormone levels, due to the constant vibration of the ground from a continuous source of vehicles (Lercher 1996; Matheson et al. 2003).

### Respiratory health

Air quality is known to be poor within 200 metres of a highway, and very poor within 50 metres (Genereux et al. 2007). To accurately gauge air quality near a highway, measurements must be taken at the source of pollutants, near the highway itself. Interpolated data from a limited number of air quality stations that are located more than 1000 metres away from a highway will not give an adequate representation of air quality in close proximity to highways or neighbourhoods at ground level (Cao 2006). As a consequence, the data from a small number of air quality stations in Montreal is insufficient to estimate variations in air quality at differing distances from highways.

There is a relationship between traffic pollutants and an increased incidence of asthma and respiratory allergies (Nillson 1999; Lewis 2000; Puliafito 2002; Holguin 2007). Typically, investigators have measured distance from highways to residential areas to establish a link with hospital admission records for asthma. One major conclusion of these studies is that low-income neighbourhoods near highways tend to have higher per capita rates of hospital admission for respiratory problems (Delfino et al. 1997; Maupin 2004); and it is often suggested that this is due to the higher density of highways in low-income neighbourhoods. In addition, there are experimental studies relating the concentration of particulate matter (PM), i.e., fine airborne particles, to non-asthmatic respiratory problems (Heinrich 2004; Brunekeef 2005; Morganstern 2008).

Asthma is a serious health hazard and has been the focus of many traffic pollution studies; several have concluded that traffic exposure and asthma are related (Rosas 1998; Hoek 2001; Chang 2009). Two variables that have been used are proximity to a highway and traffic volume. Studies applying these variables, including some Montreal studies, show an increased occurrence of asthma in the elderly as well as in young children (Delfino et al. 1997; Hoek, 2001; Gehring 2002; Holguin 2007). Respiratory ailments in all age groups are higher within 200 metres of a major highway or road, with the highest rate of occurrence being reported within 50 to 75 metres from the highway (Van Vliet 1997; Nicolai 2003; Ryan et al. 2005). Infants living in close proximity to a major roadway have a higher incidence of wheezing as compared to unexposed infants (Ryan et al. 2005). Many studies have demonstrated the relationship between air pollutants and respiratory problems. A recent study measured the amount of exhaled nitrogen monoxide (NO) in humans and concluded that it was the most reliable and measurable pollutant for explaining the relationship between respiratory difficulties and traffic (Holguin et al. 2007). This is of interest because exhaled NO can be used to estimate inhaled NO and is a good predictor of traffic pollutant levels (Holguin 2005). Other studies have found that of all the air pollutants responsible for wheezing in infants, only nitrogen dioxide ($NO_2$) and NO showed significant correlations with the prevalence of asthma; as well, asthma-related emergency room visits were positively correlated to these two air pollutants (Andersen 2008; Garty et al. 1998). Additionally, $PM_{2.5}$ (extremely fine airborne particles) in automobile exhaust is considered to be the most hazardous PM because as it is inhaled it is easily absorbed in the lungs. It may be that $PM_{2.5}$ and NO are equally responsible for respiratory disorders; however there may be additional unidentified air pollutants that are also responsible for respiratory problems.

### Sleep and Quality of Life

Respiratory health problems and noise related to large roadways are significant contributors to sleep problems (Sleep in America Poll 2007). The quality of sleep is directly related to quality of life, defined as an individual's emotional, social and physical health. A lack of sleep is associated with an increase in the risk of prostate cancer, hypertension and type II diabetes (Kakizaki et al. 2008; Mehmet Birhan et al. 2008; Javaheri et al. 2008; Beihl et al. 2009; Tasali et al. 2009). It has been shown that decreased sleep in early childhood is also associated with an increased risk of obesity (Touchette et al. 2008). Several studies have demonstrated that sleep deprivation resulted in decreased mental performance and hyperactivity in children (Pilcher and Huffcuff 1996; Touchette et al. 2007). Overall, these physical and psychological factors have been shown to affect sleep and, hence, directly affect individual's emotional, social and physical health.

The effects on physical and mental health translate into direct and indirect social costs. The annual economic impact of accidents related to sleep deprivation has been estimated at 67 to 88 billion Canadian dollars (Leger 1994).[1] In addition, the economic loss from decreased work productivity and increased vulnerability to sickness in the province of Quebec was estimated at 6.5 billion Canadian dollars (Daley et al. 2008).

## Highways, Health and Socioeconomic Status

In general, scientific findings suggest that neighbourhoods do have an effect on health status, above and beyond individual socio-demographics and behavioral characteristics (Ross, Tremblay & Graham 2004; Feldman & Steptoe 2004). As neighbourhood socio-economic status increases, so does the relative benefit to health (Feldman & Steptoe 2004; Pickett & Pearl 2001). The city of Montreal has been found to be the most segregated and unequal in terms of income in Canada (Ross, Tremblay & Graham 2004). Overall, it has been predicted that urban neighbourhoods in Canada are becoming gradually more homogenous in terms of their societal and health status (Ross, Tremblay & Graham 2004). This is important because it shows that as the gap between the rich and poor increases, the impact on the health of people of low socioeconomic status will also enlarge.

Studies show that there is a distinct relationship between health, proximity to highways and socioeconomic status. This is due to the fact that people with low incomes tend to reside near highways because property is less expensive and this in turn negatively impacts their health (Genereux et al. 2007). In general, people of lower socio-economic status are more likely to be exposed to the most noise, the worst air quality, and overall poorer environmental conditions than those with higher socioeconomic status (Link & Phelan 2002; Evans & Kantrowitz 2002). Households with incomes below $10,000 have been found to have average sound exposure levels more than 10 dBA higher than households above $20,000 annual income (Evans & Kantrowitz 2002). This is coupled with the fact that short-term exposure to fine PM exacerbates existing pulmonary and cardiovascular disease; and long-term exposure increases the risk of cardiovascular disease and death (Brugge, Durant & Rioux 2007; Feldman & Steptoe 2004; Evans & Kantrowitz 2002). These studies show that in general, low-income neighbourhoods are more likely to have a higher occurrence of health issues than those individuals living in higher income neighbourhoods.

Economically disadvantaged individuals are disproportionately exposed to traffic related pollutants which have been associated with increased respiratory morbidity and mortality as they often reside along major roads (Smargiassi 2007). Several studies link socioeconomic status and traffic intensity to low birth weight in newborns (Subramanian et al 2006;

Smargiassi 2007; Zeka, Melly & Schwartz 2008). One study linking mothers' educational attainment, newborns' birth weight and distance to highway found that mothers with a lower education who lived closer to a highway were more likely to give birth to newborns with low birth weight than mothers with a higher education who reside in a similar area (Zeka, Melly & Schwartz 2008). The increased vulnerability to air and noise pollution may be due to low socioeconomic status, whereby residents are more likely to be exposed to poor living conditions, occupational hazards, poorer nutrition and psychological stress than people of a higher socioeconomic status (Zeka, Melly & Schwartz 2008). Overall, it is suggested that the socioeconomic status, and its correlation to environmental and neighbourhood quality, has a profound impact on health.

## Conclusion

The noxious chemicals and sounds coming from automobiles on high-density roads have detrimental effects on human health for those living within the close proximity of these roads. The risks associated with proximity to highway are numerous: there is an elevated risk of respiratory, mental and sleep disorders, and an overall decrease in health and quality of life. Overall, these findings have important implications for the environmental assessment and construction of road projects.

The Ministère des Transports du Québec (MTQ) proposal will place lanes of traffic in close proximity to residential neighbourhoods. The data used by the MTQ to assess the baseline air quality is inadequate due to the small number of air quality stations and their large distance from the Turcot Interchange. On average, the air quality stations are five kilometres away from the interchange itself (own calculations). Such a large-scale interpolation cannot predict higher concentrations of pollutants that are expected to occur within 200-metres of Autoroute 720, which connects to the Turcot Interchange and borders the residential neighbourhood of the *Village des Tanneries*.

## Notes

We wish to extend our thanks to Dr. Jochen Jaeger and François Thérien for their input and contribution.

1. Currency adjusted to 2009 Canadian dollars

## References

Andersen, Z. J., S. Loft, M. Ketzel, M. Stage, T. Scheike, M. N. Hermansen, and H. Bisgaard. "Ambient Air Pollution Triggers Wheezing Symptoms in Infants." *Thorax* 63, no. 8 (2008): 710-6.

Babisch, W. F., B. Beule, M. Schust, N. Kersten, and H. Ising. "Traffic Noise and Risk of Myocardial Infarction." *Epidemiology* 16, no. 1 (2005): 33-40.

Beihl, D. A., A. D. Liese, and S. M. Haffner. "Sleep Duration as a Risk Factor for Incident Type 2 Diabetes in a Multiethnic Cohort." *Annals of Epidemiology* 19, no. 5 (2009): 351-7.

Bluhm, G., E. Nordling, and N. Berglind. "Road Traffic Noise and Annoyance--an Increasing Environmental Health Problem." *Noise and Health* 6, no. 24 (2004): 43-9.

Bluhm, G. L., N. Berglind, E. Nordling, and M. Rosenlund. "Road Traffic Noise and Hypertension." *Occupational and Environmental Medicine* 64, no. 2 (2007): 122-26.

Brugge, D., J. L. Durant, and C. Rioux. "Near-Highway Pollutants in Motor Vehicle Exhaust: A Review of Epidemiologic Evidence of Cardiac and Pulmonary Health Risks." *Environmental Health* 6 (2007): 23.

Brunekreef, B., and J. Sunyer. "Asthma, Rhinitis and Air Pollution: Is Traffic to Blame?" *European Respiratory Journal* 21, no. 6 (2003): 913-15.

Cao, J., M. F. Valois, and M. S. Goldberg. "An S-Plus Function to Calculate Relative Risks and Adjusted Means for Regression Models Using Natural Splines." *Computer Methods and Programs in Biomedicine* 84, no. 1 (2006): 58-62.

Chang, J., R. J. Delfino, D. Gillen, T. Tjoa, B. Nickerson, and D. Cooper. "Repeated Respiratory Hospital Encounters among Children with Asthma and Residential Proximity to Traffic." *Occupational and Environmental Medicine* 66, no. 2 (2009): 90-8.

Daley, M., C. M. Morin, M. LeBlanc, J. P. Gregoire, and J. Savard. "The Economic Burden of Insomnia: Direct and Indirect Costs for Individuals with Insomnia Syndrome, Insomnia Symptoms, and Good Sleepers." *Sleep* 32, no. 1 (2009): 55-64.

Delfino, R. J., A. M. Murphy-Moulton, R. T. Burnett, J. R. Brook, and M. R. Becklake. "Effects of Air Pollution on Emergency Room Visits for Respiratory Illnesses in Montreal, Quebec." *American Journal of Respiratory and Critical Care Medicine* 155, no. 2 (1997): 568-76.

Evans, G. W., and E. Kantrowitz. "Socioeconomic Status and Health: The Potential Role of Environmental Risk Exposure." *Annual Review of Public Health* 23 (2002): 303-31.

Feldman, P. J., and A. Steptoe. "How Neighbourhoods and Physical Functioning Are Related: The Roles of Neighbourhood Socioeconomic Status, Perceived Neighbourhood Strain, and Individual Health Risk Factors." *Annals of Behavioral Medicine* 27, no. 2 (2004): 91-9.

Garty, B. Z., E. Kosman, E. Ganor, V. Berger, L. Garty, T. Wietzen, Y. Waisman, M. Mimouni, and Y. Waisel. "Emergency Room Visits of Asthmatic Children, Relation to Air Pollution, Weather, and Airborne Allergens." *Annals of Allergy, Asthma and Immunology* 81, no. 6 (1998): 563-70.

Gehring, U., J. Cyrys, G. Sedlmeir, B. Brunekreef, T. Bellander, P. Fischer, C. P. Bauer, D. Reinhardt, H. E. Wichmann, and J. Heinrich. "Traffic-Related Air Pollution and Respiratory Health During the First 2 Yrs of Life." *European Respiration Journal* 19, no. 4 (2002): 690-8.

Genereux, M., N. Auger, M. Goneau, and M. Daniel. "Neighbourhood Socioeconomic Status, Maternal Education and Adverse Birth Outcomes among Mothers Living near Highways." *Journal of Epidemiology and Community Health* 62, no. 8 (2008): 695-700.

Heinrich, J., and H. E. Wichmann. "Traffic Related Pollutants in Europe and Their Effect on Allergic Disease." *Current Opinions Allergy Clinical Immunology* 4 (2004): 341-48.

Hoek, G., Meliefste, K., Brauer, M., van Vliet, P., Brunekreef, B., and Fischer, P. "Risk Assessment of Exposure to Traffic-Related Air Pollution for the Development of Inhalant Allergy, Asthma and Other Chronic Respiratory Conditions in Children *(TRAPCA)*". Utrecht: *IRAS University*, 2001.

Holguin, F., S. Flores., Z. Ross, M. Cortez M Molina, L. Molina, C. Rincon, M. Jerret, K. Berhane, A. Granados, and I. Romieu "Traffic-related exposures, airway function, inflammation, and respiratory symptoms in children." *European Respiratory Journal,* 21, no. 6, (2007): 913-5.

Javaheri, S., A. Storfer-Isser, C. L. Rosen, and S. Redline "Sleep Quality and Elevated Blood Pressure in Adolescents." *Circulation,* 118, (2008): 1034-40.

Kakizaki M., K. Inoue, S. Kuriyama, T. Sone, K. Matsuda-Ohmori, N. Nakaya, S. Fukudo, and I. Tsuji. "Sleep duration and the risk of prostate cancer: the Ohsaki Cohort Study." *British Journal of Cancer 99* no. 1 (2008): 176-8.

Leger D. "The cost of sleep-related accidents: a report for the National Commission on Sleep Disorders Research." *Sleep, 17,* (1994): 84-93.

Lercher, P. "Environmental noise and health: An integrated research perspective." *Environment International* 22, no. 1 (1996): 117-129.

Lewis, S.A., J. M. Corden, G. E. Forster, and M. Newlands. "Combined effects of aerobiological pollutants, chemical pollutants and meteorological conditions on asthma admissions and A & E attendances in Derbyshire UK, 1993–96." *Clinical & Experimental Allergy,* 30, no. 12 (2000): 1724-32.

Link, B.G. and J.O. Pheland. "McKeown and the idea that social conditions are fundamental causes of disease." *Health, Policy and Ethics,* 92 no.5 (2002): 730-732.

Matheson, M., and S. Stansfeld. "Noise pollution: non-auditory effects on health." *British Medical Bulletin* 68 (2003): 243-257.

Maupin P., and P. Apparicio. "Relationships between Ambrosia artemisiifolia sites and the physical and social environments of Montreal (Canada)." *Geoscience and Remote Sensing Symposium , IGARSS 04. Proceedings. International 1* (2004): 238.

Maupin, P., and A. L. Jousselme, "Vagueness, a multifaceted concept – a case study on Ambrosia artemisiifolia predictive cartography." *Geoscience and Remote Sensing Symposium. IGARSS 04 Proceedings. IEEE International 1* (2004): 363.

Mehmet Birhan, Y. Kenan, T. Okan Onur, Y. Ahmet, Ý. Oguzhan, B. Gokhan, and T. Izzet "Sleep quality among relatively younger patients with initial diagnosis of hypertension: Dippers versus non-dippers." *Blood Pressure*, 16 no. 2 (2008): 101-5.

Morgenstern, V., A. Zutavern, J. Cyrys, I. Brockow, S. Koletzko, U. Krämer, H. Behrendt, O. Herbarth, A. von Berg, C. P. Bauer, H. E. Wichmann, and J. Heinrich, "Diseases, allergic sensitization, and exposure to traffic-related air pollution in children." *American Journal of Respiratory Critical Care Medicine*, 177, no. 12, (2004): 1331-7.

Nicolai, T., D. Carr, S. K. Weiland, H. Duhme, O. von Ehrenstein, C Wagner, and E von Mutius. "Urban traffic and pollutant exposure related to respiratory outcomes and atopy in a large sample of children." *European Respiratory Journal* 21, no. 6, (2003): 956-63.

Nilsson, L., O. Castor, O. Löfman, A. Magnusson, and N.I.M. Kjellman, "Allergic disease in teenagers in relation to urban or rural residence at various stages of childhood." *Allergy 54*, (1999): 716-21.

Muller-Wenk, R. "A method to include in LCA Road Traffic noise and its health effects." *International Journal of Life Cycle Assessment* 9, no. 2 (2004): 76-85.

Passcher, W., and W. Passcher-Vermeer. "Noise exposure and public health." *Environmental Health Perspectives* 108 (March 2000): 123-131.

Papadakaki, M., T. Kontogiannis, G. Tzamalouka, C. Darviri, and J. Chliaoutakis. "Exploring the effects of lifestyle, sleep factors and driving behaviors on sleep-related road risk: A study of Greek drivers." *Department Accident Analysis and Prevention*, 40, (2008): 2029-36.

Pickett, K. E., and M. Pearl. "Multilevel Analysis of neighbourhood outcomes: A Critical Review." *Journal of Epidemiology Community Health* 55, (2001): 111-122.

Pilcher J.J., and A. I. Huffcutt. "Effects of sleep deprivation on performance. A meta-analysis." *Sleep, 19*, (1996): 318-26.

Puliafito, E., M. Guevara, and C. Puliafito. "Characterization of urban air quality using GIS as a management system." *Instituto para el Estudio del Medio Ambiente (IEMA), Universidad de Mendoza*, (2002).

Ross, N. A., S. Tremblay, and K. Graham. "Neighbourhood influences on health in Montreal, Canada." *Social Science and Medicine*, 59, (2004): 1485-1494.

Rosas, I., H. A. McCartney, R. W. Payne, C. Calderon, J. Lacey, R. Chapela and S. Ruiz-Velazco. "Analysis of the relationships between environmental factors (allergens, air pollution, and weather) and asthma emergency admissions to a hospital in Mexico City." *Allergy*, 53, (1998): 394-401.

Smargiassi, A., K. Berrada, I. Fortier, and T. Kosatsky, "Traffic intensity, dwelling value, and hospital admissions for respiratory disease among the elderly in

Montreal (Canada): a case-control analysis." *Journal of Epidemiology, Community and Health*, 60, (2006): 507-512.

Stansfield, S., M. Haines, and B. Brown. "Noise and health in the urban environment." *Review of Environmental Health* 15, no. 1-2 (2000): 43-82.

Subramanian, S.V., J. T. Chen, D. H. Rehknopf, P. D. Waterman, and N. Krieger "Comparing Individual and area based socioeconomic measure for the survelliance of health disparities: A multi-level analysis of Massachusetts Births, 1989-1991." *American Journal of Epidemiology*, 164, no. 9, (2006):823-834.

Tasali, E., R. Leproult, and K. Spiegel. "Reduced Sleep Duration or Quality: Relationships with insulin resistance and type 2 diabetes." *Progress in Cardiovascular Diseases,* 51, no. 5 (2009): 381-391.

Touchette, E., D. Petit, J. R. Séguin, M Boivin, R E Tremblay, and J Y. Montplaisir, "Associations between sleep duration patterns and behavioral/cognitive functioning at school entry." *Sleep* 30 no. 9 (2007): 1213-19.

Touchette E., D. Petit, R. E. Tremblay, M. Boivin, B. Falissard, C. Genolini, and J. Y. Montplaisir. "Associations between sleep duration patterns and overweight/obesity at age 6." *Sleep 31* no.11 (2008): 1507-14.

Van Vliet, P., M Knape, J. de Hartog, N. Janssen, H. Harssema, and B. Brunekreef, B. "Motor vehicle exhaust and chronic respiratory symptoms in children living near motorways." *Environmental Respiratory Journal,* 74, (1997): 122-32.

Washington State Department of Transportation. "Environment—Air, Acoustics and Energy." (2009), http://www.wsdot.wa.gov/Environment/Air/Traffic-Noise.htm.

WB&A Market research for the National Sleep Foundation. Sleep in America Poll, (2007).

Zeka, A., S. J. Melly, and J. Schwartz. "The Effect of socioeconomic status and indices of physical environment on reduced birth weight and preterm birth in Eastern Massachusetts." *Environmental Health* 7, no. 60, (2008): 1-15.

# CHAPTER NINE

# Trucking and the Turcot: Balancing Urban Quality of Life with the Economic Imperatives of Truck-based Freight Transportation

*Jacob Larsen*

Proponents of the current proposal to rebuild the Turcot and adjoining interchanges assert that maintaining high capacity for moving transport trucks is essential to the health of the Quebec and Montreal economies. It is beyond debate that transportation of goods is an essential component of Montreal's economic vitality and that the vast majority of goods are transported by truck. However, the assumption that economic prosperity for the region is contingent on unfettered truck access to the Turcot Interchange remains dubious at best. The growth of trucking as the predominant mode of goods transport in the twentieth century has been linked to environmental degradation and negative health impacts. Critics of truck transport argue for shifting freight transportation to rail; however, the decentralized nature of global transportation renders this decision totally unrealistic; such a policy change is also beyond the scope of local decision-makers. While the negative effects of trucking should be mitigated when possible, it is fanciful to imagine a freight transportation network that does not include trucks. Truck-based goods transport has become an essential component to our modern way of life. If an alternative plan for the Turcot Interchange to that of Transport Quebec (MTQ) is to be considered, it must recognize the necessity of transport trucks to the economic well-being of cities and make strategic provisions their presence in the Montreal region.

This chapter takes a pragmatic approach to freight transport by suggesting solutions that could form part of a comprehensive plan for trucking in the Greater Montreal Area. Specifically, by developing a strategy for diverting truck traffic away from the centrally-located Champlain Bridge, the proportion of trucks on the Turcot Interchange could decrease dramatically.

Reducing truck traffic on the Turcot Interchange should parallel the modal shift from passenger vehicles to public transport called for in other chapters of this book; however, it is important to remember that passenger and freight transport remain two distinct branches of transportation, serving different purposes. Evidence of the separateness of these domains is especially apparent in Western Europe, where freight transport by truck has been increasing steadily over the past 30 years while public transit ridership in European cities is the envy of every North American transportation planner.[1] While sharing the same road space, automobile and transport truck traffic each warrant separate solutions, tailored to their particular characteristics.

In the case of Montreal, shifts towards public transport together with a strategic plan for truck-based freight transport would make a downscaled interchange a feasible option. Many recommendations have already been made concerning freight transportation in Montreal in recent years; this chapter will highlight some relevant findings from the Nicolet Commission (2003) and draw on precedents from other regions that could be successfully applied to satisfy Montreal's current and future goods transportation needs. Some possible solutions explored include bypass highways such as the long-awaited Autoroute 30; truck-only lanes on existing routes; and the prioritization of truck traffic across other bridge spans. Careful implementation of some of these measures could result in an abatement of the truck traffic using the Turcot Interchange, effectively relieving the need to rebuild the complex in the same monumental proportions.

**A brief history of transport in Montreal**

Montreal's importance as a regional centre in the continental network of truck transport can be traced back to its growth in the nineteenth century, and to the development of two other types of infrastructure: water and rail. The construction of the Lachine Canal in 1825, allowing vessels to circumvent the dangerous Lachine rapids and providing controlled hydropower for emerging industries, is often cited as a pivotal moment in Montreal's industrial history. In terms of water-borne transportation, though, the deepening of the St. Lawrence River downstream from Montreal in 1850 and again in 1880 had a greater impact on regional transport, making Montreal an important port for ocean-going vessels and contributing to its development as a major trading centre.[2]

However, Montreal's predominance in the region was only guaranteed by the completion of the Victoria Bridge across the St. Lawrence River in 1859, a private undertaking providing Montreal firms with year-round access to its closest year-round ice-free port in Portland, Maine. Montreal's metropolitan status was assured at that point, as competing railways companies such as the Canadian Pacific Railway and Canadian Northern Railway invested massively in the city to compete in the expanding market created by the Grand Trunk Railway (GTR).

The transition from rail to road as the primary mode for freight transport occurred in parallel to the twentieth century rise of the automobile and can be traced back to a few decisive infrastructural projects. From a policy perspective, the passage of the Trans-Canada Highway Act (1949) and the National Interstate and Defense Highways Act (1956) in the United States prompted an important modal shift in North American goods transportation. The 1950s also marked the advent of containerized freight and the completion of the St. Lawrence Seaway, which has been linked to Montreal's diminished importance as a shipping hub as ocean-going vessels gained access to markets located farther west in the Great Lakes basin.[3]

Many critics have accused governments of subsidizing the trucking industry through road infrastructure while privatizing railways and allowing them to be swamped by increasing operating costs. Another explanation for the growth of trucking that should be noted was the steady decrease in real cost for liquid fossil fuel that occurred through the twentieth century; between 1918 and 2002 the real cost of gasoline decreased from $3.50 per gallon to $1.50 (in 2008 dollars) before beginning its steep increase in the past decade.[4]

While many explanations have been given for the growth of the trucking industry, the shift is probably best explained at the global scale by macroeconomic trends; truck transport has become the natural choice in the context of decentralized centres of production and consumption and highly specialized markets.

### Montreal's strategic location

Located between the key markets of Ontario, the Canadian Maritime provinces, and the Northeast United States, Montreal's location has ensured its importance as a hub in the intercontinental trucking industry. While data on trucking in Quebec are scarce, the MTQ reports that there are 152,000 weekly truck trips entering or leaving the territory of the Montreal metropolitan area, 21% of which are in transit to achieve trade between other regions of Quebec or North America.[5]

For these trucks in transit through Montreal, the absence of a southern bypass route makes the Champlain Bridge the most logical link, necessitating heavy truck use of the Turcot Interchange. Of the remaining weekly truck traffic, the same study reports that 68% have their origin or destination on the Island of Montreal, with the St. Laurent industrial park comprising the largest generator of trips; while there is a lack of reliable data on trucking on the Island of Montreal, the MTQ estimates that the

**Figure 9.1** Trajectory of Autoroute 30 with Montreal highways

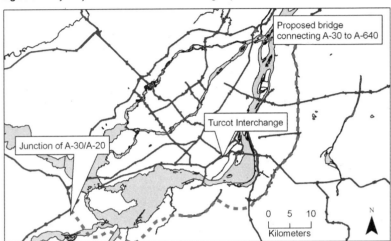

St. Laurent industrial park contains 85,000 jobs related to trucking, almost double that of any other area on the island.[6]

Trucks traveling between the St. Laurent industrial park and the South Shore towards destinations in southern Quebec, the Maritimes or the Northeast United States are also most likely to opt for the Champlain Bridge, crossing the Turcot Interchange en route. As shown in **Figure 9.1**, the strategic location of the Champlain Bridge serves as a magnet in the region, attracting high volumes of truck traffic to the Turcot Interchange. The bottleneck of truck and car traffic currently produced in the Champlain Bridge corridor is central to the MTQ's rationale for rebuilding the Turcot Interchange according to the same specifications. However, there are alternatives; **Figure 9.2** shows an alternative route using the Mercier Bridge, which passes fewer residential areas and could reduce the negative health impacts of truck transport. The two alternatives—the Champlain and Mercier routes—have prompted reflection on other occasions.

### The Nicolet Commission

The *Commission de consultation sur l'amelioration de la mobilité entre Montréal et la Rive-Sud*, hereafter referred to as the Nicolet Commission, was created in January 2001 to study the possibility of building a new infrastructure (bridge or tunnel), in partnership with the private sector, to solve the

**Figure 9.2** St. Lawrence River Basin with residential land use. Champlain Bridge and proposed Mercier Bridge priority truck routes are identified

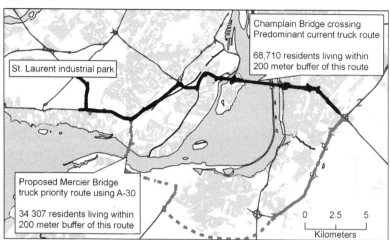

St. Laurent industrial park

Champlain Bridge crossing
Predominant current truck route

68,710 residents living within
200 meter buffer of this route

Proposed Mercier Bridge
truck priority route using A-30

34 307 residents living within
200 meter buffer of this route

0       2.5       5
Kilometers

Source: 2006 Census

chronic problem of congestion on the bridges between Montreal and the South Shore. Central to the Commission's *raison d'être* was the growing congestion of the Champlain Bridge, which according to the projections used by the Commission would result in a 14% increase in average travel time.[7]

However, in spite of its original directives, the Commission quickly became a forum for debating the entire problem of mobility between Montreal and the South Shore. In part due to the divergent viewpoints between advocates of new infrastructure and the advocates of public transit, the focus of the Commission was expanded to related issues: management of demand, infrastructure funding, greenhouse gas emissions, quality of life and public health.[8]

The Commission also devoted a chapter of its final report to the process of consultation in transportation infrastructure development in Quebec, insisting that the discussion of options and alternatives should take place upstream of the decision-making process. This openness to problem-definition exemplified by the Nicolet Commission stands in stark contrast to the project development process pursued with respect to rebuilding the Turcot and to other recent infrastructure projects in the province of Quebec.[9]

The focus of the Commission's final report remained largely on its original mandate concerning specific infrastructure projects. While warning against induced demand, the Commission recommended two additional lanes for road vehicles and two additional lanes for public transit on the Champlain Corridor, the busiest span across the St. Lawrence River. The Commission also recommended widening the Honoré-Mercier Bridge for two additional lanes for truck transport and public transit and a new bridge connecting the Autoroutes 640 and 30 at the eastern tip of the island, completing a long-awaited bypass route that would allow many trucks to avoid crossing onto the island entirely.

Finally, the Commission recommended a comprehensive plan to manage demand for road space in the city's downtown core in conjunction with improved public transit, reducing congestion across the river spans through a mode shift. Despite its comprehensive approach and stated goals of increasing public transit, some environmental groups criticized the Nicolet Commission for not going far enough and questioned the privileged consideration of infrastructure projects originating from the private sector.

To date, the Nicolet Commission remains the most comprehensive study of the transportation issues facing Montreal and the South Shore and suggests promising alternatives for dealing with the present challenge surrounding the Turcot Interchange. Not surprisingly, independent transportation experts in Montreal have drawn on the Commission's conclusions extensively. The Groupe de recherche urbaine Hochelaga-Maisonneuve (GRUHM) cites the Commission's estimate of a 3% annual increase in truck traffic and argues that an enlarged Mercier Bridge connected to the Autoroute 30 could become the primary truck crossing for the western portion of the island; completing the Autoroute 30 without prioritizing the Mercier Bridge for trucks would result in a 30% increase of trucks on the Champlain Bridge by 2021, meaning over two million additional trucks crossing the Turcot Interchange annually.[10]

Moreover, simply completing the Autoroute 30 bypass would likely result in little if any reduction in trucking on the island; 68% of trips currently originate or end on the island. As revealed in **Figure 9.1**, the Mercier Bridge provides easy access to the St. Laurent industrial parks and, as mentioned above, has a lower concentration of residential areas nearby, making it the optimal span on the western portion of the island to handle a larger load of truck traffic, provided that consultations with neighbouring Kahnewake community result in a mutually acceptable plan.

Elsewhere in North America, other jurisdictions have successfully adopted strategies to deal with the flow of heavy trucks through their territory. Truck-only lanes and differentiated tolls are two such strategies. Truck-only lanes vary in scale and function, providing either access to a particular facility such as a port or an alternate route around an urban area; they have been used in Southern California and are planned in New Jersey.[11] Provision of such lanes on the Mercier Bridge would act as an incentive for truckers to use that span and effectively direct them away from the Champlain Bridge.

Another study found that differentiated tolls are more effective than dedicated lanes in managing truck flows.[12] Time-of-day pricing is one method of differentiated tolls, charging drivers more when crossing in peak periods; this has been implemented at several major access points to Manhattan Island with promising results.[13] Applied to trucks crossing the Island of Montreal, this disincentive would charge drivers a fee crossing the Champlain Bridge that would not apply to the Mercier Bridge crossing. However, as already noted by Nicolet Commission, the level of congestion that is already found on these two spans would require additional truck facilities on the Mercier Bridge to handle this additional flow. In other words, the "stick" may not produce the desired results unless the "carrot" is made available.

### A comprehensive vision for the Turcot Interchange

To describe the Turcot Interchange as a giant knot requiring meticulous disentanglement is an appropriate metaphor for the transportation challenges facing Montreal today. From the air, the many tendrils of its swooping overpasses conjure up such an image, however it also points to the persistent and calculated measures that will be needed to craft a more sustainable transportation strategy for the region.

This chapter argues that a viable alternative to the MTQ's proposal for the Turcot Interchange is possible and presents several options for truck transport. Drawing on the findings of the Nicolet Commission and proposals by the GRUHM, as well as precedents from other regions, this chapter sketches a few possible alternatives, such as bypass highways around dense urban areas, truck-priority routes and differentiated pricing. It has been repeatedly stressed that opponents to the MTQ's proposal must face the reality of a continental freight transportation system that is dominated by trucks. Truck transport has become an essential component in our economic system; failure to recognize this basic fact will result in a fatal loss of credibility to any alternative proposed.

As a subset of this larger web of issues surrounding the Turcot Interchange, the transport of goods by truck in the Montreal region warrants consideration on its own merits and requires a distinct plan to be managed effectively. At present, the MTQ has developed no such plan, opting instead for a *laissez faire* approach to road-based freight transportation, with sporadic, reactionary responses to demand for roads through massive infrastructural investments.

Where the Turcot Interchange is concerned, the MTQ proposes to rebuild a similar structure to accommodate even greater volumes of traffic, while providing no policy directives to guide planning for truck transport in the region. Until such a strategy is defined and implemented, the negative impacts associated with heavy truck traffic will continue to increase.

### Conclusion

This chapter represents one attempt to articulate an alternative vision to that proposed by the MTQ by emphasizing minimizing the negative impacts of trucking. The Minister of Transport will require further study of this issue, involving the collection and analysis of more precise data related to trucking, in order to arrive at an optimal solution.

Ultimately, though, the central transportation issue facing Montreal today is not a question of infrastructure or information but one of leadership; at present, it appears that the governing party has decided that it is not advantageous to invest its political capital in bringing about a paradigm shift towards sustainable transportation. If progress towards this end is to be achieved, it will require that all levels of government invest both economically and politically. Indeed, it will be up to the public to demand that they do so.

### Notes

1. U.S. Department of Transportation Federal Highway Administration. "Freight Management and Operations: International Freight Studies" (1999). http://ops.fhwa.dot.gov/freight/freight_analysis/euro_scan/index.htm.
2. David Hanna. "The Importance of Transportation Infrastructure" In *Montreal Metropolis: 1880-1930*, edited by Isabelle Gournay France, 45-57. Toronto: Stoddart, Vanlaethem, 1998.
3. Brian Slack. "Gateway or Cul-de-Sac the Saint Lawrence River and Eastern Canadian Container Traffic." *Etudes Canadiennes 26*, (1989), 49–58.
4. U.S. Energy Information Administration. "CPI-U Inflation index". Cited by U.S. Bureau of Labor Statistics, http://www.bls.gov.

5. Quebec Ministry of Transportation. "Interurban Heavy Vehicle Traffic in Québec—1999 Trucking Survey." Government of Quebec, 1999. http://www.mtq.gouv.qc.ca/portal/page/portal/entreprises_en/camionnage/camionnage_international/enquete_camionnage_1999

6. Le Comité interrégional pour le transport des marchandises and Transport Quebec. "Diagnostic sur la congestion routière et le transport des merchandises", 1999. http://www.citm-transport.org/pdf/Congestion-diagnostic.pdf.

7. Quebec Ministry of Transportation. "Mieux se déplacer entre Montréal et la Rive-Sud." Commission de consultation sur l'amélioration de la mobilité entre Montréal et la Rive-Sud, (2003). Quebec City.

8. Mario Gauthier, Laurent Lepage, Louis Simard and Valérie Saint-Amant. "Environmental Assessment of Road Infrastructure: Toward the Development of a Regional Framework for Integrating Climate Change Factors." Canadian Environmental Assessment Agency, CEEA Research and Development Monograph Series, 2002. http://www.ceaa.gc.ca/default.asp?lang=En&n=4DB25320-1&offset=16&toc=show.

9. Ibid.

10. Pierre Brisset (presentation). "Analyse du Transport des merchandises dans le cadre du Portrait et Diagnostic du Plan de Transport de Montréal". (2004) http://gruhm.org. Accessed May 7, 2009.

11. Michael J. Fischer, Dike N. Ahanotu and Janine W. Waliszewski. "Planning Truck-Only Lanes: Emerging Lessons from the Southern California Experience." *Transportation Research Record,* no. 1833. (2003), 73-78.

12. André de Palmaa, Moez Kilanib, and Robin Lindsey. "An Economic Analysis of Segregating Cars and Trucks". *Journal of Urban Economics.* 64: 2, (2008). 340-361.

13. Kaan Ozbay, Ozlem Yanmaz-Tuzel, José Holguín-Veras. "Impacts of Time-of-Day Pricing on Travel Behavior: General Findings from Port Authority of New York and New Jersey Initiative". *Transportation Research Record 1960* (2006), 48-56.

# CHAPTER TEN

# Transportation and Highways in Montreal

*Ian Lockwood and Joel Mann*

## Introduction

As Montreal rethinks its 1960s freeway infrastructure, it faces daunting problems: congestion, the continued viability of its transit system, ever-growing demands on revenue streams, impending economic concerns (e.g., fuel prices, manufacturing exportation, etc.), and increasing environmental concerns (e.g., climate change, air quality, etc.) seem insurmountable at times. The patterns of investment, growth, and development that have sustained the City for the past fifty years are showing their age and some are diminishing. Bold cities, elsewhere, have adopted strategies very different from what Montreal is doing. Only a new approach to transportation investment will ensure future prosperity for Montreal. This chapter provides an overview of how today's transportation strategy came to be, how well it has delivered, and the major issues that could and should cause it to change in the future.

## Context of the Montreal's Transportation Plans

In developing transport plans for Montreal, it is important first to understand what problems the city faces and what opportunities it has for moving forward. Any new projects will represent major investments that should promote civic enhancement and economic growth. Montreal was undeniably one of the success stories of North America, a city that drew on a legacy of ingenuity and culture to develop as a thriving centre of commerce with national and international influence. How Montreal will adapt its transportation system to catalyze what should be a growing and prosperous city involves understanding the components of urban transport and access.

## The Role of Transportation

The average person would rather not think about or deal with traffic. When given a choice, people prefer to focus their lives on family, work, and leisure activities. Transport should simply be a means to accomplish these ends efficiently. When this relationship falls out of balance, however, the quality of people's lives lowers. To many residents of North American cities, such an imbalance has occurred. The past 30 years have seen a substantial increase in commute times. The cost of transportation as a percentage of income for families has grown, too.

Public space devoted to transportation allows people to access their different activities and services, and facilitates trade; it is the binding element in our cities. In the past half-century, the solution has been to provide ever-larger and faster highways and arterial streets, resulting in ever-longer commutes and sprawling suburban development, transferring value and people out of Montreal. It is increasingly clear, however, that this model eventually reaches limits beyond which it cannot or should not be sustained. In Montreal, those limits involve the road congestion that stifles economic growth and community health, with many side effects. The high costs of such long trips, the health impacts of more sedentary lifestyles, and the deteriorating air quality in Montreal all suggest that a reconsideration of the conventional transportation paradigm is timely.

It is worth considering lessons from other cities that have sustained longer and larger growth. Many of these cities have demonstrated the open-mindedness required to allow economic growth while developing models for well-connected, easy-to-navigate streets, and options for travel. Montreal is not the first city to need to change course.

The time is right for Montreal to pause and reconsider the wisdom of automobile-oriented, conventional highway strategies and to consider improving quality of life and pursuing sustainability. Rather than seeking to move more cars further and at higher speeds, Montreal should strive to move more people shorter distances. Instead of believing that Montreal can "solve" congestion, the City should focus on managing its congestion. Now is Montreal's opportunity to create a better model.

### Transport as a Place-Making Element

Conventional transportation professionals claim that highways and highway widenings are built in response to development needs. They do not recognize that transportation infrastructure is, rather, the primary determinant of what future growth will occur. The presence or absence of transportation infrastructure is a key element that allows the extent, location, type, and mix of development. For example, Montreal's 1960s freeway construction program facilitated high speed driving out of town and fueled the suburban growth that continues even today. Clearly, transportation investments were a primary cause of that development. Examples of land use responses can be seen in both suburban and urban contexts. Most newer green-field places that were built with conventional approaches to transportation infrastructure (highways connected to large arterial streets connected to collector streets connected to local streets connected to driveways) fostered development forms such as single story shopping centres with surface parking, disconnected office and industrial parks, tract housing, and largely separated land uses. However, suburban transportation and development forms represent the communities of "choice" for many, despite the tradeoffs: low density, disconnected, development coupled with wide, fast streets that discourage walking, biking, and transit. Such suburban communities were never meant to sustain very high densities. Growth in such places is limited by their inability to accommodate more cars.

A different model and level of growth can be observed when infrastructure is built into the street network as part of the urban form. Well-connected networks of streets provide superior access, efficient trip distribution, and multiple routing options. They are also better suited for walking, cycling, and transit use and so better able to support greater development intensities than the suburban patterns. A fine-grain street network creates smaller blocks that are better for pedestrians.

Walkable streets are better for more intense development and are a basic requirement for frequent and reliable public transport service. Walkable, bikeable, and public transport-oriented places have much higher "people-moving" capacities than places that rely primarily on single-occupant automobiles. Thus, connected, multimodal transportation strategies are critical to accommodate dense development. Montreal has many examples of these successful strategies in its older areas with robust networks of streets and land uses characterized by taller structures and buildings with doors that address the streets, such as Downtown and Old Montreal. The city needs to employ such urban strategies today.

Streets are also the primary public spaces in the city. Unlike their European and Latin American counterparts, North American cities largely abandoned their historic legacy of place-making through networks, squares, and plazas after World War II. The post-war function of streets became to move people and goods from place to place as quickly as possible. Consequently, streets do not fulfill the fundamental role of place-making, leaving a void of arenas for public life. Our cities need to reassert their role in bringing people together and fostering civic life. The character and the care given to a community through the provision of contributing streets is evident to any visitor immediately, although this may not be readily apparent to residents. It is a simple matter of looking at and experiencing the transportation infrastructure of the city and comparing it to other places. Naturally, land uses respond to such care over time as well. Michigan Avenue (Chicago), Newbury Street (Boston) and Broadway (New York) are important transportation thoroughfares, but they also accommodate and foster a balance of civic and economic functions that generally reflect the character of their cities through these signature streets. Likewise, Montreal's investments indicate its character and care for its community.

### Fiscal Importance: Transportation as an Investment

While the role of transport as an element of place-making is important, its role as an economic investment should not be underestimated. Even in 1998, before the steady rise in fuel prices, the average household spent about a fifth of its income on transportation-related expenses, an amount equal to the combined total amount spent on health care and food.

While people may accept these as the cost of living in a modern society, Montreal should strive to do better. If Montreal were to invest differently, could it reduce these costs? Could it generate greater tax revenues from development? Could it reduce air pollution and provide more opportunities for physical activity? Could it lower its carbon footprint and dependence on oil? Could it make the city more attractive and valuable?

For much of the 20th century, conventional transportation theories assumed that new highway infrastructure was the key to fostering economic growth. The working premise was that congestion, created by these highways and resulting sprawl, could be remedied by adding more car-capacity (i.e., more lanes and more highways).

Following this paradigm fueled the rapid growth of the suburbs in the Montreal region at the expense of the city. Like cities and regions across North America, Montreal found that its newly added car-carrying capacity was soon exhausted, leading to a long-term commitment to construct more and more highway infrastructure, in a vicious circle. In addition to the high capital costs, maintenance was expensive: something that has been done badly in Montreal over the last several decades. Consequently, Montreal should think twice before accepting even more highway infrastructure.

## Older Cities and Change

Refurbishing of old city centres is a recent phenomenon that is occurring across North America, spurred by different approaches to transportation and development in those cities and also peoples' changing attitudes and values. People increasingly want to be part of the action, living where they do not have to 'drive to town' for a great quality of life. This is in keeping with the fundamental purpose of cities which has not changed since cities were invented: cities exist primarily for the purpose of maximizing exchange (i.e., to bring people together for the efficient exchange of goods, services, knowledge, ideas, labor, capital, entertainment, resources, etc.)

Cities and metro areas also exist to minimize transportation. The less people have to move themselves or their goods, the better and more efficient the system is. The role of transportation in cities is about access, not speed or mobility. Good access translates into everything from a good block structure, to an interconnected street system, to on-street parking, to walkability, to proximity of complementary land uses, to being able to use multiple modes of transport. Conventional transportation theories ignore the structural aspects of access and its relationship to land use because they consider transportation *on its own*, as if the city's structure and land uses were unrelated.

Since the rise of North American suburban development in the second half of the 20th century, suburbs have appeared more attractive than central cities due to their newer infrastructure, seemingly lower costs, new schools and freedom from crowding, crime, and health concerns. Post-war policies, infrastructure design and funding mechanisms supported these perceptions. Yet the realities of suburban expansion suggest that they are not sustainable on many levels. Now that they are aging, the suburbs are undergoing the same challenges as the cities. Moreover, they face additional problems due to their inefficient infrastructure patterns that are both expensive and hard to change. These problems are compounded by declining property values, changing demographic patterns, health challenges, energy issues, and an increased demand for public services.

In addition, the notion of the suburbs providing freedom from traffic congestion has largely been discredited. To a great extent, these problems can be traced to the layout: their 'dendritic' street systems. Disconnected street patterns, large blocks, and limited points of neighbourhood access were implemented by land developers to discourage non-local traffic and to provide a sense of tranquility and enclosure based on notions of space, quiet, and privacy. This was fine from the perspective of any single development but, when added up, the sum was unsuccessful. At some point, a relatively long trip from the neighbourhood to everywhere else was necessary, and disconnected street networks funneled the traffic loads onto overburdened arterial streets and highways. In turn, the arterial and suburban interchanges attracted concentrations of shopping and businesses that required access and further hampered the model. Hence, many of the most congested corridors became those serving relatively low density suburban communities.

### The 'New' Importance of Place

Recent studies, particularly those by Richard Florida, suggest that cities that will succeed in the future will be those that can retain professionals who work in innovative fields. These cities will not only have stronger economic foundations, but they will also have greater social diversity and higher levels of quality of place. Montreal could be enjoying high levels of re-population and renewed investment if it were to provide the basic physical and civic infrastructure to support it. This includes a balanced transportation system that offers modal and routing options and allows the streets and transit to support and celebrate an urban environment. Highway building does the opposite.

The city's streets are the vantage point from which most people experience the city. Streets contribute greatly to the perception and image of the city as a place where people choose to be. Streets, combined with parks and squares to fill all the spaces between the buildings, create the "public realm." They provide access and support to the land uses as well as connections between those land uses. All of these factors combine to attract and retain good people.

### Examples of Practices from Other North American Cities

Montreal is not alone in contemplating a future that looks very different from its past. North American cities have witnessed broad demographic and attitudinal changes in the last two generations. This trend shows no signs of abating and the cities that embrace these changes will have competitive advantages over other cities. Montreal's current highway building, steeped in the values of thirty years ago, will not help the city compete.

This section describes the experience of four other North American cities that are hailed for combining land use and transportation planning. Each has encountered issues similar to those facing Montreal, and each has addressed these issues in its own way. They provide Montreal with guidance in its transportation strategy.

### Charlotte, North Carolina

Charlotte, North Carolina, is a city with a bright future. It is a major financial and distribution centre of the Southeastern United States. The city has prospered through the strength of its business community and by pro-development attitudes and policies. Charlotte's unified approach to planning for growth and new infrastructure is a key to its success. The City has aligned its different departments and responsibilities and, in doing so, has streamlined planning and development. The alignment is a simple yet powerful concept that Charlotte calls its Centres, Corridors and Wedges Growth Strategy. The idea is that development intensity be tied to the areas where the city's infrastructure can support it. Charlotte has created this link by focusing the investment of public resources along corridors with centres of compact development and a broad palette of land uses, while reserving the remaining 'wedge' areas for open space and less intense development.

Unlike other cities with strong legislative frameworks for growth management, Charlotte does not have a comprehensive plan tying together the missions of its different departments. In the absence of such a unifying element, the growth framework organized around centres, corridors and wedges is crucial. The framework facilitates the alignment of departmental interests so that growth is guided to areas that can support it and steered away from areas that cannot. This also allows planning to respond to changing community values as each department's needs and understanding of community concerns are reiterated through joint planning processes.

Charlotte has identified five primary growth corridors, linear districts with concentrations of high-capacity transportation facilities that include street networks and transit infrastructure. The City began operating its first rail transit service in the first of five corridors in November 2007 and is planning similar changes for the remaining four. The land use envisioned for these corridors is a mix of moderate to high-density residential, office, retail, industrial, and warehouse/distribution uses. This conforms to Charlotte's current and growing role as a transportation and distribution centre in the Southeast while acknowledging its strong economic growth and demand for housing.

Transportation planning supports the need for connectivity and walkability within the growth areas. The city concentrates transportation funding for street and network projects in the centre and corridor growth areas.

It also identifies key walking, cycling, and urban livability components associated with larger transit projects.

### Vancouver, British Columbia

Vancouver's planning policies promote a high quality of urban livability. It has broadened its transportation system to promote non-automobile modes of transportation, responding to increasingly scarce resources that could never support the auto-dominated alternative. As a result, Vancouver has become a world leader in sustainability, forward-thinking urban development, and general quality of life. *The Economist* and the World Health Organization have proclaimed Vancouver as a model of high urban livability.

Yet perhaps even more notable has been its coordination of regional interests and the development of a well-integrated policy framework to guide development and growth and to link land use and transportation. A unified transportation authority, the South Coast British Columbia Transportation Authority (locally referred to as TransLink), is responsible for road construction and maintenance and transit infrastructure and operations.

TransLink asserts a large role in regional coordination between transportation and land use. It has promoted public transit as a primary travel mode, mainly as a result of policies that do not base project decisions on added capacity for automobiles: it has tied investment in new transit infrastructure to regional land use planning and development. As a result, it has more control over project prioritization between transit, roadway, and bicycle/pedestrian projects and greater flexibility to use funding among all of them. This single transportation entity supports the region's policy of prioritizing transit investment. This has allowed the region and municipalities to plan transportation infrastructure in a uniform and coordinated manner.

One of TransLink's main strategies in developing a balanced transportation system has been the concept of legible transit (i.e., a transit system that the public understands and trusts). In moving to establish a sense of permanence and legibility, TransLink has found that transit should be seen as public infrastructure and not merely a service. Montreal could learn a great deal from Vancouver. Interestingly, there are no highways in the City of Vancouver.

### Chicago, Illinois

Though their size and historical growth are different, Chicago and Montreal are similar. They are the economic and cultural centres of their regions and have broad, diverse economic bases rooted in transportation and industry. They rose to prominence through transportation infrastructure and continue to play an important distribution function within national and international economies.

Chicago has a particularly strong focus on rail transport, being served by all of the North American Class I railroads. It has an additional 14 small railroads with 2,800 track-miles of rail (excluding rail yards), 500 freight and 700 passenger trains per day and 37,000 freight car and 20,000 intermodal movements per day.

At the same time, Chicago has become a leader, in recent years, in promoting and adopting environmentally-conscious and sustainable growth and development. It has earned attention for the green roof on its City Hall, new bicycle lanes, and improved sidewalks. Chicago sees the enhancement and maintenance of public infrastructure for the safety and convenience of all users as fundamental to a city whose citizens can enjoy a high quality of life without depleting natural resources.

Nearly 23 per cent of its land area is public right-of-way and Chicago believes the land should do more than just move traffic around—it should also contribute to Chicago's sense of place. Chicago has adopted a Complete Streets Policy recognizing the right of the public to make transportation and project decisions. Chicago also seeks to utilize its rights-of-way to achieve the greatest community benefit possible. This, in turn, has led to several programs such as the Streetscape Program, which has built beautiful streetscapes, bicycle lanes, trails, improved crossings, and sidewalks.

These programs have been augmented by revisions of a Landscape Ordinance that defines standards for planting and landscaping. The Green Alley program modernized the city's 1,900 miles of service alleys with permeable surfaces that facilitate drainage, allow natural percolation to lessen the impact on the city's storm sewers, and reduce heat islands through the use of lighter surface materials.

Detroit, Michigan

During the first 50 years of the last century, Detroit's population rose from 286,000 to 1,850,000, the fourth largest city in the USA. During the second 50 years, the population shrank to 951,000 and is even lower today. Poor transportation planning was one factor that encouraged and enabled the depopulation, and included the construction of several highways, the resulting loss of street connectivity, the over-widening of several city streets, the elimination public transportation, and the abandonment of public rights-of-way to create superblocks. The result is a disconnected, barrier-ridden, pedestrian-hostile, and empty city fraught with problems of disinvestment and abandonment. Several "silver bullet" attempts to enliven the City through large office complexes and stadiums investments have failed.

Detroit was far better off before it was retrofitted with highways designed to take both people and investment away from the city. Discussion of removing highways, or parts thereof, has started but it means a shift in thinking that is difficult for transportation officials in "Motor City" to make.

It is interesting that, 50 years ago, Detroit was a wealthy and highly educated city on the cutting edge of transportation thinking. Detroit did a far more thorough job of embracing the highway vision than any other city. With 20/20 hindsight we can now compare how well that did, compared with Vancouver's strategy. Interestingly, many of the arguments that are being used today to increase highway construction in Montreal are the same as the ones in Detroit years ago, which have since been rejected by an increasing number of cities.

## Related Practices and North American Trends

### Highway Patterns

Many North American cities added highways in the 1950s, 1960s, and 1970s with similar patterns of decay and disinvestment as Montreal and Detroit. Several cities recognized the destructive patterns and abandoned subsequent highway projects, including South Pasadena, San Francisco, Philadelphia, and Toronto. The areas where the highways were supposed to go were "saved" and are still nice places today. More recently, cities are beginning to remove their highways and, without exception, the social and economic effects have been very positive, as in New York City, Chattanooga, Portland, San Francisco, and Milwaukee. More cities are working to remove their downtown highways as part of their revitalization and sustainability efforts, as is occurring in Trenton, Syracuse, Toronto, Hartford, and Seattle.

It is no coincidence that in 1950, at the height of Detroit's success, the City had a well-connected, complete, and highway-free street network. The ensuing road widenings and highways, which were rammed through the heart of the City, became conduits for people and investment to leave. Unless many of these mistakes are corrected Detroit will never recover. A similar future awaits Montreal; it needs to embrace place, people, and access, not highways, automobiles, and speed.

Key Highway Principles

1. Context-sensitive street designs in cities, and especially in downtowns, should foster social and economic exchange.

2. Downtown streets that share traffic loads tend to succeed; dendritic road hierarchies that include highways try to concentrate traffic loads and fail.

3. Barriers in downtowns should be removed, not buried. Cities are about being connected and this should be reflected in the street network.

4. Transportation systems, serving downtowns, should be multimodal and encourage greener modes of transportation.

5. Visitors should feel that they are "connected" to the "place." Roads designed for highway speeds detract from "place." Streets in cities should be "complete" and enforce safe speeds through design.

Limiting Factors

For decades, highway critics have hoped that some limiting factor would slow, stop, or reverse highway building in cities. Energy, pollution, climate change, health, quality of life, disinvestment, population flight, and automobile dependence were all candidates but failed to change the paradigm. Ministries of Transportation seem interested in sustainable practices, working with stakeholders, and cooperating with municipal jurisdictions only when they are compelled to by a more important limiting factor or, in rare situations (e.g., San Francisco and Toronto), by extraordinary leadership or community pressure. The New Jersey and Pennsylvania Departments of Transportation would still be battling congestion with expensive road widenings and highways had they not run out of money. Money has become the important limiting factor. When transportation money is plentiful, then conventional, highway-oriented, interests are good at getting the money and spending it.

Just because there is highway money for Montreal, does not mean that highways are good for Montreal.

Hierarchy of Roads

For the last 50 years, the transportation industry has had an unfortunate preoccupation with the hierarchy of roads (local, collector, arterial, and highway), in a dendritic form, even though it is abundantly clear that cities and metro areas are better off with interconnected street networks that provide parallel and redundant routing choices, inform land use patterns to a human scale, and spread the transportation loads.

The conventional, dendritic, practices result in i) poorly connected streets which disadvantage all modes, except the automobile; ii) super blocks scaled to accommodate automobiles; and iii) the concentration of transportation on a few routes (making those streets too big and dangerous for pedestrians, cyclists, and transit).

### VMT and Fuel Use

Conventional transportation practices reward increases in vehicle-miles-traveled (VMT) and fuel consumption with federal and provincial money to fund road widening projects, exacerbating the problems. However, the solution is not to throw away such measures upon which to base such rewards. The solution is to change the practice 180 degrees. Places that have projects and programs that reduce their VMT and fuel use should get funding and those that increase these measures should get nothing. In this way, efficiency is rewarded. Both VMT per person and fuel use per person needs to be considered along with total VMT and fuel use. This is absolutely crucial to the future of cities and the future of Montreal.

### Speed and Cities

It is a very modern and ill-conceived idea that big roads must be fast roads. Yet speed on big roads has spread cities out, increasing trip lengths, reducing density, and reducing exchange.

Speed, by definition, is anti-city, anti-metro area, inefficient, and automobile-oriented. However, the issue remains unaddressed and misunderstood in Montreal. Complete streets catering to all modes require slow, safer speeds which allow proper block structures, access, and exchange to occur. Networks work better with slower traffic. Slower speeds, by design, are good for cities.

The argument for faster roads relies on the idea of time "savings" and related measures of effectiveness including, among others, delay, travel time and average speed. The reasoning is that saving travel time is good in cities, faster roads save time, and therefore faster roads are good.

However, they are only good if all else remains equal. Unfortunately, the behavior and land use responses to faster roads more than undoes any benefit derived from changes to make the road faster. *In city after city, including Montreal, congestion worsened after vast sums were spent in an attempt to speed up the roads through highways, widenings, smoothing curves, and so forth.* The conventional paradigm never fully considered speed's effects on promoting use of the automobile and discouraging walking, cycling, and transit.

## Speed and Capacity

There is a difference between speed and capacity (for motor vehicles). Some of the best addresses in European cities are on high capacity streets. There is no reason why all city streets cannot be walkable, bicycle-friendly, business-friendly and transit-oriented as well. If they are fast, they only benefit the automobile.

Montreal should reject high speed highways and arterials. The related effects are so highly damaging that high speed roadways should be replaced with city-friendly designs or removed entirely. Ironically, when highways in cities are being used the most, during peak hours, motorists suffer the slowest and most frustrating travel experiences. Cities are about access and highways are about speed; they are like oil and water.

Montreal needs to keep in mind that high speed highways in cities are a very recent idea. Now we know that they simply violate the context of the city. Where they have been introduced, they have done harm; where they have come out, good things happen; where they were halted, good things continued. The pattern is clear.

## Distribution of Goods

Goods movement and freight are very important in metro areas. Jonathan Barnett's research found that "The de facto land use plan of the United States has been the interstate system." Urban sprawl followed the highways out of cities because cheap land was developed in automobile-dependant ways and the system consistently clogged up. Retailers opened 'big box' stores to exploit this "free" resource. Their trucks delivered huge quantities of produce and, due to their purchasing power and low distribution costs, they sold it for less than their smaller competitors who internalized the distribution costs. The rub here is that shoppers drove long distances along the same highways to buy cheap goods. So we see that society underwrites big stores with the most expensive sort of infrastructure and the worst transportation mode available.

This model is wrong from every perspective of city-making. We must consider the benefits of shorter trips for both goods and local production; growing or making products for the city and metro markets in closer proximity. As long as we build costly highways to keep costs down (for the truckers and the foreign producers who benefit the most), we should not be surprised that one of the biggest North American exports is jobs to places with cheap labor and/or tax structures.

Though buying cheap foreign products may seem appealing, North Americans are subsidizing some else's distribution costs with our transportation money. By placing our industries at a disadvantage, we have fewer jobs and less money to buy things. We have developed a system that harms our domestic industries and workers.

### Leveraging Transit for Development

Some American cities have begun using high-frequency (premium) transit to encourage development that increases transit ridership and adds value to cities through increased population and tax base, as in Charlotte and Vancouver. Improved transit infrastructure reduces the need to accommodate automobiles in new development. This frees land and lowers development costs that can be applied to increasing a development's yield. Consider St. Louis' 25-year transit modernization plan which is expected to generate a $2.3 billion return in business sales.

### Facilitating Healthy and Active Lifestyles

Many cities have invested in projects that improve lifestyles. Austin, Texas has a plan that assures "that the development of the urban environment is compatible with the unique natural and constructed features of the Austin area." This has allowed Austin to realize high population and job growth in recent years.

The successful cities will be those that create engaging open space and pedestrian environments. In most North American cities, streets account for 20% to 25% of the total land area and many cities are taking initiatives to shape this space into forms that provide greater benefits to the public than moving cars. Denver's 2007 Downtown Area Plan calls for boulevards to be a backbone of downtown economic development. Chicago has taken similar steps to construct streets that improve drainage, reduce heat and lessen the need for street lighting.

### Land Use Options

Some people believe that the best transportation policy is a good land use policy that recognizes the need for a broad range of housing options. This refers to levels of affordability, size of dwelling units, proximity to grocery stores and recreational centres, employment, and other daily activities. The idea is to support the shorter trip and provide travel mode options. After decades of building big houses and dwelling units, cities recognize a need for smaller homes. Public transit and sidewalks are especially important as people age.

There is a recognized pattern of people moving to the suburbs in search of single-use zoning, recreational complexes, large grocery stores, and "good" schools, though all in a very resource-hungry and automobile-dependant way. Historically, public incentives and policies promoted the migration of people as well as investment. The sprawling growth was developer-driven and they were not required to provide a street network. The focus was on building and selling "product" (developer-speak for homes, strip malls, etc.) and constructing as little infrastructure as feasible in order to maximize profits. This meant designing for automobiles and exploiting the old farm-to-market roads as well as the later installed highways (e.g., Autoroute 25 in Montreal).

Similarly, there is the disinvestment and less than efficient use of land in many of the deserted inner city areas and downtowns, especially where highways are planned (e.g., along the Notre Dame corridor). But, ironically it is these places that have the infrastructure and network structure potential to be efficient and to foster exchange. So, instead of disadvantaging these areas with highways (or a near-highway in the case of Notre Dame) and subsidizing suburban houses with wider, faster roads, why not spend that money on older areas and refurbish inner city streets, schools, parks, grocery stores, and housing?

Leverage the cultural institutions, the history, the main streets, and the inherent locational advantages of these areas. Catering to suburban commuters is both costly and ineffective. Cities must, instead, discourage undesirable trends and support the structure and the places with the most potential.

# Afterword

Montreal is indeed at a crossroads. Poised to embark on some of the largest infrastructure projects in decades, costing billions of dollars, decision-makers at the Ministère des Transports du Québec are about to determine the future structure of the City.

Impending decisions will bring Montreal closer to one of two destinies. One, car-inspired, is likely to lead to more urban sprawl, more pollution and a weaker downtown. The alternative, transit-oriented approach will produce a more compact city, a significant net decrease in Montreal's "carbon footprint," and fewer negative health effects for residents.

No mistake should be made at this point: What happens with the Turcot will structure the City for decades to come. Future generations will judge which decisions were well taken and which ones were mistakes.

The history of transportation as it affects city growth has occurred in three phases: from the beginning of time until roughly 1870, the city was built for walking and the use of horses. It was compact and human-scaled. In the second phase, with the introduction of the train and the tramway, cities stretched out tendrils, with the population clustering along it like sugar crystals, especially at intersections and close to train stations, on new main streets. But with the invention of the automobile, the city has started to sprawl, in particular since 1950.

An evocative French phrase captures this affect: *la ville éparpillée*—the Scattered City.

Montreal is a beautiful city and there is much to be proud of. It has some of the most famous inner-city neighbourhoods in North America: Côte-des-Neiges, the "chic Plateau" and Mile-End, now so famous for its hipsters and lively music scene. But Montreal also sprawls, and it sprawls in all directions.

The automobile's tarred companion, the road, requires massive public subsidy to build and maintain. Every year, some of the most valuable farm land in North America is sterilized by asphalt, as big-box stores, parking lots and new sprawling subdivisions push the limits of our City farther and farther outwards.

This is the curse of the Third phase of City development.

Of course, the suburbs are not all bad. The City of Montreal continues to lose some of its population to the North and South shores, and to Laval, our sister island. Every year, thousands of residents migrate to the suburbs, taking the tax base to the edges of the metropolis. Families are drawn there by the quiet lifestyle and an affordable house with a safe yard for their children to play. And many of them are cursed with a daily commute that costs them an average 64 minutes a day (the Canadian average, according to a recent study). And interminable traffic jams.

Yet, the 21$^{st}$ century city will be a compact city. The initiation of the Fourth phase of City building is just around the corner. In fact, it has already begun in several cities, as the chapters in this book have highlighted.

The contributions collected in this volume have made it very clear that Montreal is facing crucial choices: we must make up our minds now about what our city will look like in the coming decades. Montreal cannot wait for the electric car, that mirage of the Detroit Auto Salon. Montreal cannot wait the twenty or thirty years—or fifty years—it will take to significantly alter the proportion of electric vehicles within Quebec's automobile and truck fleet. We need to take long-term steps now.

The city and its inhabitants should not sit idly by, while a decision on the Turcot is made for them. Montreal needs to act.

"Cities are not passive participants in the processes that affect resource availability, waste absorption, input and output markets. They also accumulate the very technologies, institutions, people and knowledge that may help identify and implement solutions. In the long run, the ability to reduce their ecological, and especially their greenhouse gas footprint, will be key to slowing, halting or even reverting global climate change. The longer cities wait to put in place strategies to mitigate greenhouse gas emissions, the more they will have sunk investments in infrastructures, technologies and institutions that maintain or increase emissions of greenhouse gases, and the greater climate change they will need to adapt to."[1]

Montreal has talents that rival the best in emerging green technologies. Montreal companies design trains and buses and tramways and these are built in Quebec. Montreal has the creativity to become a world leader and a world exporter of solutions for our environmental crisis. Several authors in this book have provided persuasive alternatives to the renovation of the Turcot Interchange. They demonstrate that Montreal has a great potential to set an example.

The world is watching for clear steps. When a city decides to renovate its aging car infrastructure, few pay attention, and those that do, usually scowl.

But when a city decides to scrap its aging car infrastructure, the whole world sits up and pays attention. This is a city, they say, that is truly addressing the problem of global warming. This is a city that cares.

As a prime example, Copenhagen has embarked on a dramatic effort to make itself the Climate Capital of the world. The City aims for 20% less carbon dioxide emissions by 2015. In 2015, 50% of the people who work or study in Copenhagen are to use the bicycle as their means of transportation to and from work and studies—the current percentage is 36 %. To control pollution, only busses and trucks with particle filters will be allowed into the downtown "environmental zone".

San Francisco, perhaps the most "Montreal-like" city in the United States, is systematically removing highways from the inner city. After an earthquake damaged the elevated Embarcadero freeway, separating the City from a swath of waterfront, it was never repaired. It was finally rebuilt as a true urban boulevard, blending a pedestrian promenade, a bike corridor and a new tramline. Plaques mark the spots where the pillars once stood.

Even Detroit, the "Motor City", is discussing the removal of some of its highways in favour of a more pedestrian-scaled approach.

The world's cities will be well served by embarking on the next phase of city development, one that prioritizes mass transit, healthy environments, and liveable neighbourhoods that are compact and family friendly. The city of the 21$^{st}$ century will not pave over local farmland, but foster and protect that land, and feed itself from that land.

The Montreal of the 21$^{st}$ Century must be a City that truly belongs to that century. We must look forward, not backward, when we determine the fate of the Turcot Interchange. The Turcot provides Montreal with a unique window of opportunity for bringing about the changes that the city of the future so badly needs.

*Pierre Gauthier*
*Jochen Jaeger*
*Jason Prince*

**Note**

1. *Competitive cities in a changing climate: Introductory issue paper.* http://www. oecd.org/dataoecd/22/40/41446908.pdf, retrieved 02 March 2009.

# Postface

À n'en pas douter, Montréal est à la croisée des chemins. Prêts à s'engager au cours des prochaines décennies dans quelques grands projets d'infrastructure qui coûtent des milliards de dollars, les décideurs du ministère des Transports du Québec sont sur le point d'établir les termes de l'évolution spatiale de la Ville pour le futur.

Une alternative s'offre à Montréal. Des décisions imminentes engageront ce dernier, dans l'une ou l'autre voie. La première, inspirée par l'automobile, entraînera vraisemblablement la ville vers une expansion tentaculaire, un niveau de pollution accru et la perte de vitesse de son centre-ville. La voie de rechange, une approche axée sur le transport en commun, engendrera une ville moins distendue et plus dense, entraînera une importante diminution du bilan carbone de Montréal, et aura moins d'effets négatifs sur la santé de ses résidants.

Ne laissons pas place à l'erreur : ce qui émanera du projet de reconstruction de l'échangeur Turcot façonnera la ville pour des décennies à venir. Les générations futures jugeront leur héritage à l'aune de nos erreurs et des bonnes décisions que nous prenons aujourd'hui pour elles.

L'histoire du transport en rapport avec la croissance de la ville est marquée par trois étapes : depuis l'aube des temps jusqu'à environ 1870, la ville était conçue pour la marche et l'utilisation des chevaux. Elle était dense et à l'échelle humaine. La seconde étape a connu l'arrivée du train et du tramway, les villes ont étendu leurs tentacules, la population s'y agglutinant comme des abeilles, particulièrement aux intersections et près des gares, sur les nouvelles rues principales. Mais l'invention de l'automobile a provoqué l'expansion incontrôlée de la ville, notamment depuis les années 1950.

Une expression traduit bien l'état qui caractérise la troisième période : la ville *éparpillée*.

Montréal est une ville magnifique dont nous avons toutes les raisons d'être fiers. Ses quartiers centraux figurent parmi les plus renommés en Amérique du Nord : Côte-des-Neiges, le « chic » Plateau et le Mile-End. C'est une ville bouillonnant d'activité en son centre, branchée, créative et dont la réputation est solidement établie sur le plan artistique. Mais Montréal connaît aussi une expansion désordonnée. Elle s'étale dans toutes les directions.

Alliée goudronnée de l'automobile, la route, nécessite l'investissement d'énormes ressources publiques pour sa construction et son entretien.

Chaque année, l'asphalte rend stérile quantité de ce qui compte parmi les meilleures terres agricoles de l'Amérique du Nord pour leur substituer des centres commerciaux et leurs stationnements ainsi que de nouveaux lotissements tentaculaires qui repoussent les limites de notre ville de plus en plus loin.

C'est le fléau de la troisième période de l'histoire de la mobilité dans la ville.

Bien sûr, les banlieues ne sont pas en tous points mauvaises. La Ville de Montréal continue de perdre une partie de sa population au profit de la rive nord et de la rive sud et de Laval, l'île-sœur. Chaque année, des milliers de résidants migrent vers les banlieues, élargissant l'assiette fiscale des confins de la métropole. Les familles y sont attirées par un mode de vie paisible et des maisons abordables dotées d'une cour sécuritaire où leurs enfants pourront jouer. Mais un bon nombre de ces banlieusards est obligé de faire la navette quotidiennement, au prix de 64 minutes par jour (la moyenne canadienne, selon une récente étude) ou plus. Cela sans compter d'interminables congestions routières dont la fréquence est en augmentation.

Il y a espoir malgré tout. Il y a tout lieu de croire que la ville du XXIᵉ siècle sera une ville dense. La quatrième période est à nos portes. En fait, elle a déjà débuté dans plusieurs villes, comme on l'a souligné dans ce livre.

Les articles réunis dans ce volume ont clairement démontré que Montréal est à la croisée des chemins. Confrontés à un choix crucial, nous devons décider maintenant ce que sera le Montréal du XXIᵉ siècle. Montréal ne peut pas attendre l'arrivée de la voiture électrique, ce mirage du Salon de l'auto de Détroit. Montréal ne peut pas attendre pendant les quelque vingt ou trente ans — voire cinquante ans — requis pour qu'une proportion significative du parc automobile et de camions du Québec soit constituée de véhicules électriques. Nous devons prendre maintenant des mesures qui nous engagent à long terme.

La ville et ses habitants ne devraient pas se montrer indifférents alors qu'on s'apprête peut-être à prendre une décision à leur place à l'égard du projet Turcot et de projets autoroutiers similaires. Montréal doit bouger.

[*Traduction*]

*Les villes ne doivent pas rester passives devant les processus qui ont un effet sur la disponibilité des ressources, l'absorption des déchets, le marché des intrants et des extrants. Elles accumulent également les technologies mêmes, les institutions, les personnes et le savoir qui sont susceptibles d'aider à déterminer et à mettre en œuvre des solutions. À la longue, la capacité de réduire leur empreinte écologique, notamment celle des émissions de gaz à effet de serre constituera la clé pour ralentir, interrompre ou même renverser le changement climatique planétaire. Plus les villes attendent pour mettre en place des stratégies visant à réduire les émissions de gaz à effet de serre, plus elles auront à assumer des investissements dans les infrastructures, les technologies et les institutions qui maintiennent ou augmentent les émissions de gaz à effet de serre, et devront s'adapter à un plus grand changement climatique.[1]*

Montréal possède des compétences qui rivalisent avec celles des meilleures villes en matière de technologies vertes émergentes. Les entreprises montréalaises assurent le développement, le design et la mise en marché de trains, d'autobus et de tramways qui sont fabriqués au Québec. Montréal possède la créativité pour devenir un leader mondial et un exportateur de solutions destinées à remédier à la crise environnementale. Plusieurs auteurs ont suggéré dans ce livre des solutions de rechange convaincantes pour la rénovation de l'échangeur Turcot. Ils ont contribué de ce fait à faire la preuve que Montréal dispose d'un potentiel considérable pour devenir une référence en matière de développement soutenable.

Le monde s'attend à des mesures concrètes. Lorsqu'une ville décide de rénover ses infrastructures vieillissantes destinées aux automobiles, peu de gens s'y intéressent, et ceux qui le font, se renfrognent habituellement.

Mais lorsqu'une ville décide de mettre la hache dans son infrastructure autoroutière pour se tourner vers le transport collectif, le monde entier s'éveille et s'intéresse aux solutions retenues. C'est une ville, dira-t-on, qui s'attaque réellement au problème du réchauffement de la planète. C'est une ville soucieuse de l'environnement et un modèle duquel s'inspirer.

À titre d'exemple remarquable, Copenhague a déployé de grands efforts pour devenir la Capitale mondiale de la lutte aux changements climatiques. La ville vise à diminuer de 20 % de ses émissions de dioxyde de carbone d'ici 2015. En 2015, 50 % des personnes qui travaillent ou étudient à Copenhague utiliseront leurs vélos comme moyen de transport pour se rendre à leur travail ou à leur établissement d'enseignement et en revenir— le pourcentage actuel est de 36 %. Afin de contrôler la pollution, seuls les autobus et les camions dotés de filtres de nouvelle génération auront droit de circuler dans la « zone environnementale » du centre-ville.

San Francisco, l'une des villes américaines qui présentent le plus de ressemblances avec Montréal, retire systématiquement les autoroutes de sont centre-ville. L'autoroute aérienne Embarcadero, séparant la ville d'une portion de sa rive et qui fut endommagée à la suite d'un tremblement de terre, ne fut jamais réparée. La décision fut plutôt prise de la démanteler pour la remplacer par un véritable boulevard urbain, où se côtoient une promenade piétonnière, une piste cyclable et une nouvelle voie ferrée. Des plaques indiquent aujourd'hui les endroits où se trouvaient les piliers. Elles constituent la vague réminiscence d'une ère que personne ne regrette. Même Détroit, la « Ville de l'automobile », songe à démanteler certaines de ses autoroutes afin de favoriser une approche plus axée sur le piéton. Les villes du monde seront qui seront les mieux servies sont celles qui s'engageront dans l'ère de l'aménagement urbain qui priorise le transport en commun, les environnements sains, et les quartiers agréables à vivre qui sont denses, conviviaux et bienveillants pour les familles. La ville du XXIe siècle ne pavera pas les terres agricoles, mais protègera la terre de laquelle elle tirera sa nourriture.

Le Montréal du XXIe siècle doit être de ce siècle. Nous devons tourner nos regards vers l'avenir et non pas vers le passé, au moment de déterminer le sort de l'échangeur Turcot. Ce projet d'infrastructure de transports procure à Montréal une occasion unique d'engager les changements essentiels à l'édification d'un futur meilleur pour notre ville.

*Pierre Gauthier*
*Jochen Jaeger*
*Jason Prince*

### Note

1. *Competitive cities in a changing climate: Introductory issue paper.* http://www. oecd.org/dataoecd/22/40/41446908.pdf, extrait le 2 mars 2009.

# La réfection de l'échangeur Turcot : une occasion unique de revoir nos façons de faire en matière d'urbanisme et de mobilité durable—et de réparer et re-développer le Sud-Ouest![1]

Les grands projets d'infrastructure sont par définition structurants, surtout ceux de réaménagement ou de remplacement en milieu urbain qui présentent des occasions uniques de planification stratégique pour le redéveloppement des secteurs concernés. Ils sont actuellement très prisés par les gouvernements sinon pas les prêteurs, pressés de rassurer la population devant l'effondrement de l'économie et des viaducs. Ils devraient cependant être aussi l'occasion de créer de la richesse par une meilleure intégration à la trame et au tissu urbains, par l'amélioration du cadre et de la qualité de vie et par une hausse des valeurs foncières et fiscales. L'optimisation des effets positifs possibles entraîne aussi des économies d'échelle et une réduction des coûts et des impacts négatifs.

Ce serait le cas ici des projets de réfection de la rue Notre-Dame et de l'échangeur Turcot (1,5 milliard chacun), aux deux extrémités de l'axe Ville-Marie, qui auront un impact majeur sur le centre-ville, mais pourraient aussi avoir des effets positifs importants sur ces deux entrées de ville en déshérence, indignes de Montréal, en les dégageant en particulier des obstacles industrialo ferroviaires désuets qui les plombent, et en dégageant au contraire les potentiels immobiliers latents considérables liés à leur situation exceptionnelle.

C'est le cas aussi de cette autre entrée de ville, l'autoroute Bonaventure, dont la reconfiguration, dans la foulée de la réouverture du canal de Lachine, s'annonce prometteuse à cet égard. D'autres avantages majeurs sont en jeu, comme l'image de ville internationale que veut se voir Montréal, mais qui passe par l'innovation urbaine comme architecturale. Ville de « création » et de design, grand centre universitaire et pôle d'expertise international en ingénierie, en transport sur rail en particulier, ville reconnue pour sa qualité de vie et la vitalité de ses quartiers centraux, métropole du premier état de l'est américain pour la réduction des GES, Montréal dispose de tous les atouts pour devenir *la* référence continentale en matière de développement urbain durable, surtout par le rôle central que les transports collectifs urbains et interurbains y sont appelés à jouer. Nous en sommes encore loin si l'on en juge par un dossier comme celui de la réfection de l'échangeur Turcot. La Ville de Montréal a déjà fait connaître son insatisfaction du projet mené par le MTQ, tant pour l'ignorance de son Plan de Transport et de la priorité donnée au transport collectif à l'échelle régionale, que pour les impacts locaux d'un rabaissement sur talus et l'absence totale d'intégration urbaine : « les mêmes paramètres, la même plomberie qu'il y a 40 ans » ! Ce sont là les trois principaux enjeux sur lesquels il faut revenir.

## Transport

Au plan du transport d'abord, contrairement au Plan de la Ville et à ses propres objectifs, le MTQ non seulement maintient, mais augmente la capacité de débit de l'échangeur par l'adoucissement des courbes et l'ajout d'accotements de chaque côté des chaussées—un concept imposé par les normes rurales de la Transcanadienne! Aucune considération pour le transport en commun sinon la possibilité d'ajouter une voie réservée aux autobus. Aucune distinction entre les circulations transitaire et pendulaire, cette dernière pourtant mieux assurée par transport collectif, en commençant par l'augmentation de la desserte actuelle par train ou par métro. Aucune mention enfin des projets de navette aéroportuaire ou du tramway de Lachine en particulier, qui permettraient de réduire cette circulation, et partant, les dimensions de l'intervention.

## Structure

Au plan de la structure ensuite, le rabaissement de la chaussée sur remblais permet peut-être de réduire les coûts de construction et les craintes ravivées d'effondrement, mais cette solution, normale en milieu rural, entraîne des impacts locaux majeurs en milieu urbain : tassement et inégalité de la chaussée, mais surtout triplement de l'emprise au sol, avec démolitions et perte importante de valeur immobilière; dangerosité accrue par proximité des polluants, bruits, poussières, et de l'accès piéton aux talus—présentés comme des attraits, des occasions de verdissement; création d'enclaves et de cul de sacs non sécuritaires et asociaux; banalisation généralisée du paysage urbain. À cet égard, il faut rappeler l'attention accordée partout ailleurs et depuis toujours aux grands ouvrages d'art vus comme des sculptures environnementales, surtout dans des lieux aussi touristiquement significatifs que les entrées de ville. Ce fut même le cas ici lors de l'inauguration de l'échangeur, loué pour son « effet cathédrale »!

## Intégration urbaine

Enfin, au plan de l'intégration urbaine, rien n'est proposé non seulement pour réduire ces impacts négatifs latéraux, mais surtout pour tenter d'optimiser les effets positifs possibles de tels grands travaux. On a ici une occasion unique de revoir et corriger les erreurs et effets négatifs du passé et d'améliorer l'environnement urbain de secteurs entiers, ici de tout l'arrondissement Sud-Ouest. La conversion du canal de Lachine, qui a fondé ces quartiers, devrait servir d'exemple pour le Grand Tronc. En effet, depuis la désindustrialisation, ce dernier ne fait que traverser et plomber ces quartiers. Pourtant de simples déviations en emprises existantes permettraient de les recoudre et de les relancer, comme celle déjà proposée à Pointe-Saint-Charles par simple déviation via l'axe Butler, et maintenant celle de Saint-Henri, seule possible en déviant les voies par l'ancien triage du secteur industriel Cabot. Pourrait-on même rêver, à long terme, à une déviation du trafic marchandise interrégional, qui n'a plus de desserte locale, via le pont du CP par exemple?

## Autres potentiels

Plusieurs autres projets potentiels sont reliés au chantier Turcot et devraient être étudiés et planifiés avec lui : la navette aéroportuaire et le tram-train de Lachine (1847); l'installation des ateliers de l'AMT prévue par elle à l'ancienne gare intermodale Turcot, mais bloquée par le MTQ pour son chantier—sans compter la possibilité d'y déménager aussi éventuellement Via Rail, libérant les terrains de grande valeur sur le fleuve après déplacement de l'autoroute Bonaventure. Un dernier projet est envisageable, celui de l'agrandissement du parc de la «falaise» Saint-Jacques par le piémont qui lui manque pour devenir utilisable et accessible, pas seulement par les 12 mètres concédés récemment pour une piste cyclable—merci pour le transport actif—, mais par prolongement de l'axe de la rue Saint-Antoine par exemple. On pourrait même y exhumer la rivière Saint-Pierre - sinon le lac à la Loutre qui avait aussi fait l'objet d'un projet plus ambitieux. Cet ancien marécage, toujours instable, se prête mal de toute façon au déplacement proposé des infrastructures lourdes de la 20 et du CN qui devraient être conservées au sud, sur la partie stabilisée de l'ancienne gare.

## Gouvernance

Tous ces projets ont déjà fait l'objet de propositions, certaines très élaborées, par divers organismes professionnels ou communautaires. Ils restent tous à être étudiés et validés, mais advenant le cas, aucun ne sera réalisable— ni aucun des problèmes appréhendés réglable—si l'on ne modifie pas le concept et l'approche du MTQ. Et aucune modification n'est possible à cet égard sans changement dans la gouvernance de ce genre de grands projets d'infrastructures en milieu urbain. Ici, si le maître d'ouvrage est bien le MTQ, la maîtrise d'oeuvre devrait normalement revenir à la Ville de Montréal, première responsable de l'urbanisme, de l'aménagement du territoire et du processus de planification conséquent. L'importance structurante d'un tel projet sur le développement ou non de plusieurs quartiers et arrondissements, comme sur la santé publique et l'avenir de la mobilité durable, devrait commander l'élaboration d'un méga *Plan particulier d'urbanisme* par une table de concertation réunissant tous les acteurs concernés, publics et parapublics, corporatifs, associatifs et citoyens.

Nous suggérons donc à la Ville de Montréal de prendre l'initiative auprès des gouvernements supérieurs pour mettre en place un tel processus démocratique de planification urbaine des grands projets. Nous suggérons également au Ministère des Affaires Municipales de s'impliquer dans cette bonification, comme au Ministère de l'Environnement d'élargir à l'ensemble de ces enjeux les questions que le BAPE doit éventuellement adresser au promoteur, le Ministère des Transports du Québec.

## Note

1. This statement prepared by the Table de Travail Turcot (TTT), presented to the public during a press conference on March 17, 2009, was supported by over 50 individuals, including urban planners, community activists, academics, geographers, ecologists and other citizens of Montreal. (For a complete list of co-signers, see    http://www.cyberpresse.ca/opinions/forums/cyberpresse/200903/24/01-839800-echangeur-turcot-faire-mieux.php).

# Appel pour la tenue d'une audience publique élargie

Montréal, le 7 mai 2009

**Madame Line Beauchamp**
Ministre du Développement durable, de l'Environnement et des Parcs
675, boul. René-Lévesque Est, 30ᵉ étage
Québec (QC)
G1R5V7

Objet : Mandat du BAPE pour la tenue d'une audience publique élargie sur les projets de transport dans l'axe de l'autoroute Est-Ouest (complexe Turcot et modernisation de la rue Notre-Dame) à Montréal

Madame Beauchamp,

En tant que Ministre du Développement Durable, de l'Environnement et des Parcs, nous vous demandons de mandater le Bureau d'audiences publiques sur l'environnement (BAPE) pour la tenue d'une audience publique élargie dans le cadre du projet de réfection du complexe Turcot sur l'ensemble des projets de transport de l'axe de l'autoroute Est-Ouest, incluant les projets suivants :
- la reconstruction du complexe Turcot,
- la navette ferroviaire vers l'aéroport Montréal-Trudeau,
- les projets de tramway et de tram-train,
- la modernisation de la rue Notre-Dame,
- les projets de bonification des trains de banlieue
- le projet Viabus.

Tous ces projets qui sont portés soit par la Ville de Montréal, le Ministère des Transports du Québec (MTQ), Aéroports de Montréal ou l'Agence métropolitaine de transport (AMT) constituent différentes solutions pour résoudre les problèmes de mobilité dans cet axe d'autoroute.

Nous sommes d'avis qu'il est inacceptable que le gouvernement tienne des consultations publiques uniquement sur le projet de reconstruction du complexe Turcot sans tenir compte de l'ensemble des projets et de leurs impacts cumulés et sans s'être doté au préalable d'une vision claire des priorités de transport et d'aménagement pour ce secteur de Montréal de concert avec la Ville de Montréal, l'AMT et Aéroports de Montréal.

En effet, dans le contexte où le gouvernement du Québec souhaite déjà investir plus de deux milliards de dollars dans la rue Notre-Dame et le complexe Turcot et qu'il ne dispose pas des fonds nécessaires pour construire tous les projets présentement à l'étude, nous devons collectivement faire des choix parmi les différentes options qui sont proposées par la Ville, le MTQ, l'AMT et Aéroports de Montréal. Ces choix devront reposer non seulement sur des critères de mobilité mais également sur des critères de mise en valeur du territoire, d'impact économique positif pour Montréal, de réduction des émissions de gaz à effet de serre (GES) et des polluants atmosphériques, de réduction de consommation d'énergie et d'espace, et d'amélioration de la qualité de vie pour les résidents des quartiers traversés.

Quels sont les projets offrant le maximum de gains en termes de mobilité, de revitalisation, de réduction des GES et d'amélioration de la santé et de la sécurité des résidants à proximité et de l'ensemble des citoyens ?

Quels sont les projets qui vont contribuer à l'atteinte des objectifs du plan de transport de Montréal, du plan de développement durable du MTQ, du plan d'urbanisme de Montréal et du cadre d'aménagement et orientations gouvernementales de la région métropolitaine de Montréal, à savoir le développement des transports collectifs et la réduction de l'utilisation de la voiture ?

Comment ces projets peuvent-ils être intégrés pour obtenir le maximum de gains avec le minimum d'investissements?

Quels sont les projets qui vont permettre au gouvernement du Québec de contribuer à la réduction des émissions de GES, de respecter son plan de réduction et d'atteindre ses objectifs pour Kyoto ?

À notre avis, toutes ces questions doivent être traitées dans le cadre d'une large consultation publique afin de pouvoir comparer les projets entre eux et nous permettre d'identifier clairement nos priorités avant même que des fonds publics ne soient affectés à l'un d'entre eux.

À cet égard, rappelons que le Vérificateur général du Québec a mentionné dans son dernier rapport que le Gouvernement du Québec investit des milliards de dollars dans le transport routier pour la région métropolitaine de Montréal sans avoir une « vision cohérente et intégrée de l'aménagement et du transport du territoire » et que la planification « nécessite une meilleure prise en compte des incidences à court et à long terme sur la santé, le contexte social, l'économie, l'environnement, l'utilisation des ressources naturelles ». De plus, le Vérificateur a clairement indiqué que le MTQ n'est pas parvenu à mettre en place un « mécanisme permanent et formel de concertation pour assurer la cohérence de l'ensemble des priorités d'intervention des acteurs de la région et établir un consensus ». Enfin, il souligne que le MTQ ne fait pas une « analyse comparée et documentée de différents scénarios possibles » et que « les investissements présentés » ne sont pas « représentatifs de la situation actuelle » [1].

Dans ce contexte, nous croyons qu'une telle audience publique élargie est essentielle pour corriger certaines des lacunes identifiées par le Vérificateur général en permettant à la population de prendre connaissance de l'ensemble des projets routiers et de transports collectifs qui sont présentement à l'étude dans l'axe de l'autoroute Est-Ouest, d'évaluer leurs impacts cumulatifs et de se prononcer sur leur pertinence. Le résultat de cette consultation publique permettra à la Ville de Montréal, à l'AMT, à Aéroports de Montréal et au MTQ d'établir une véritable vision intégrée et cohérente des transports et de l'aménagement pour l'axe Est-Ouest de Montréal.

En souhaitant une réponse positive de votre part, nous vous transmettons, Madame Beauchamp, nos plus sincères salutations.

*[signature]*

Denis Plante
Président, Conseil régional de l'environnement de Montréal

Au nom des cosignataires :

Pierre Morissette, Directeur général, Regroupement économique et social du Sud-Ouest
Yves Lévesque, Président, Coalition montréalaise des tables de quartier
Claude Lauzon, Directeur, CDEC Côte-des-Neiges/Notre-Dame-de-Grâce
Suzanne Bernard, Directrice générale, Corporation de développement communautaire de la Pointe
Michel Ducharme, Président, Conseil régional FTQ Montréal métropolitain
Gaétan Châteauneuf, Président, Conseil central du Montréal métropolitain-CSN
Steven Guilbeault, Coordonnateur général adjoint, Équiterre
Daniel Bouchard, Porte-parole, Coalition pour la réduction et l'apaisement de la circulation
Luc Gagnon, Président, Transport 2000 Québec

André Bélisle, Président-directeur général, Association québécoise de lutte contre
la pollution atmosphérique
Karel Mayrand, Directeur Québec, Fondation David Suzuki
Pierre Lussier, Directeur, Jour de la Terre Québec
Éric Darier, Directeur, Greenpeace
Alexandre Turgeon, Président, Vivre en Ville
Jérôme Normand, Directeur général, ENvironnement JEUnesse
Pascale Fleury, Coordonnatrice, Éco-quartier Pointe-Saint-Charles
Claude d'Anjou, Directrice générale, Mobiligo
Pierre Brisset, Architecte et directeur, Groupe de recherche urbaine Hochelaga-
Maisonneuve
Bernadette Brun, Directrice générale, Voyagez Futé
Owen Rose, Président, Centre d'écologie urbaine de Montréal
Pierre Ricard, Président, Concertation Ville-Émard-Côte-St-Paul
Dinu Bumbaru, Directeur des politiques, Héritage Montréal
Vincent Marchione, Président, Comité de vigilance environnementale
Patricia Clermont, Cofondatrice, Association Habitat Montréal
Sandrine Périon, Présidente, Corporation Solidarité Saint-Henri
Michel Poirier, Coordonnateur, Table de quartier du Nord-Ouest-de-l'Île de
Montréal
Nathalie Fortin, Coordonatrice, CLIC Bordeaux-Cartierville
Michel Roy, Coordonnateur, Conseil pour le développement local et commu-
nautaire d'Hochelaga-Maisonneuve
Geneviève Locas et Maggie St-George, porte-parole, Mobilisation Turcot
Genevieve Grenier, Coordonnatrice, Action-Gardien, Table de concertation
communautaire de Pointe-Saint-Charles
Marco Viviani, Directeur, développement et relations publiques, Communauto
inc.
Jonathan Théorêt, Directeur intérimaire, Groupe de recherche appliquée en
macroécologie
Marc Lessard, Président, Collectif en environnement Mercier-Est
Mireille Giroux, Coordonnatrice, Mercier-Ouest quartier en santé
Gaétan Legault, Coordonnateur, Coalition pour Humaniser Notre-Dame
Yoland Bergeron, Président, Parc Promenade Bellerive, inc.
Nathalie Berthelemy, Représentante, Comité Enviro-Pointe de Pointe-Saint-
Charles
François Saillant, Coordonnateur, Front d'action populaire en réaménagement
urbain
Christina Xydous, Organisatrice communautaire, POPIR-Comité Logement
Florence Bourdeau , Directrice, Carrefour d'aide aux nouveaux arrivants
Marie-Ève Deguire, Présidente, VRAC Environnement
Simon Racine, Directeur général, Éco-quartier Parc-Extension
Florence Bourdeau, Directrice, Carrefour d'aide aux nouveaux arrivants
Nicole Bastien, Directrice, Pro-Vert Sud-Ouest
Daniel Breton, Porte-parole, Mouvement MCN21 .

Fabienne Audette, Coordonatrice générale, Solidarité Mercier-Est
Jody Negley, Fondatrice, Comité des Citoyens du Village des Tanneries
Maude Landreville, Chargée de projet - Transport et mobilité des aînés, Table de
concertation des aînés de l'île de Montréal
Leslie Bagg, Organisatrice communautaire, Conseil Communautaire NDG
Jean Décarie, Urbaniste et membre, Table de Travail Turcot

CC. Madame Julie Boulet, Ministre des transports
Monsieur Yves Bolduc, Ministre de la santé et des services sociaux
Monsieur Raymond Bachand, Ministre des finances et Ministre du développe-
ment économique, de l'innovation et de l'exportation
Monsieur Gérald Tremblay, Maire de Montréal
Monsieur André Lavallée, Responsable du transport, ville de Montréal
Monsieur Richard Lessard, Directeur, Agence de la santé et des services sociaux
de Montréal
Monsieur Joël Gauthier, Président-directeur général,, Agence métropolitaine de
transport
Monsieur Michel Leblanc, Président et chef de la direction, Chambre de Com-
merce du Montréal Métropolitain
Monsieur Michel Labrecque, Président de la Société de transport de Montréal
Monsieur James C. Cherry, Président-directeur général, Aéroports de Montréal

## Note

1. Communiqué, 1er avril 2009, Le vérificateur général livre les résultats de la
vérification concernant la planification du transport et de l'aménagement dans la
région métropolitaine de Montréal.

# About the Contributors

**Henry Aubin** is a *Montreal Gazette* columnist. He graduated with a BA from Harvard, covered urban affairs for the *Washington Post* and has been at *The Gazette* since 1973. He has received three National Newspaper Awards. He is the author of *City for Sale* (*Les vrais propriétaires de Montréal*), a 1977 best-seller on the corporate forces behind Montreal's redevelopment, and of three other books.

**Patrick Asch :** Biologiste de la faune de formation, Patrick Asch (BSc. Agr. McGill) a développé une expertise en restauration et gestion d'écosystèmes urbains maximisant biodiversité et activités récréatives durables. Directeur d'Héritage Laurentien, Patrick coordonne des employés et bénévoles ayant réhabilité divers habitats urbains et conçus des plans de gestion écologiques, plantant ainsi 113 000 arbres et arbustes indigènes tout en accomplissant 105 000 heures de bénévolat et en sensibilisant 420 000 individus à la nature urbaine.

**Pierre Brisset** is a professional architect and member of the Order of Architects of Quebec. His current practice has involved him in many citizens' committees to resist the destructive effects of urban intrusion related to the uncontrolled proliferation of inner city highway projects. Being involved in urban integration projects, he was recently awarded an Orange Prize by Heritage Montreal for his part in restoring and recycling an abandoned commercial building for the Old Brewery Mission in Montreal.

**Erika Brown** is currently completing her Master's in Environmental Impact Assessment at Concordia University, Montreal. She received her Bachelor of Science in Environment from McGill University (2006).

**Meaghan Ferguson** is completing her Master's in Environmental Impact Assessment at Concordia University in Montreal. She has completed two Bachelors of Arts from Concordia, one in History in 2006 and one with a specialization in Human Environment in 2008.

**Raphaël Fischler** holds degrees in architecture (the Netherlands) and in city and regional planning (U.S.) and is Associate Professor in the School of Urban Planning at McGill University. He does research in the fields of land development and land-use regulation and the history and theory of planning. He is an administrator of the Ordre des urbanistes du Québec, a member of the Policy Committee of the Canadian Institute of Planners, an alternate member of the Comité ad-hoc d'architecture et d'urbanisme of the City of Montreal and a consultant with public, not-for-profit and private organisations.

**Frédéric Gagnon** completed a Bachelor of Science from the Université du Québec à Montréal (1992) in biology with a specialization in molecular biology and biotechnology. He is currently finishing his Masters in Environmental Assessment at Concordia University, and has been a professional scientific writer for the past 11 years.

**Elham Ghamoushi-Ramandi** is currently finishing her Masters in Environmental Assessment at Concordia University, she completed a BSC at the University of Windsor (2008) in Honours Physical Geography with a minor in Information Technology.

**Jeff Kenworthy**, BSc. (Hons), PhD Murdoch, Professor in Sustainable Cities, Curtin University Sustainability Policy Institute, Curtin University, Perth, Western Australia. He teaches courses and supervises postgraduate research students in the area of urban sustainability. He has 30 years experience in urban transport and land use policy with over 200 publications in the field. He is particularly noted for his international comparisons of cities around the theme of reducing automobile dependence.

**Jacob Larsen** is a Master's Candidate in Urban Planning at McGill University. As a member of Transportation Research at McGill (TRAM), his academic interests include the influence of urban form on walking and cycling, bicycle lane safety and the impact of transportation infrastructure on land use.

**Ian M. Lockwood, P.Eng.,** is a principal with the community planning firm of Glatting Jackson Kercher, Anglin Inc. in the United States. Mr. Lockwood is a Professional Engineer with Bachelor and Masters Degrees in Civil Engineering. Mr. Lockwood's work has been dedicated to evolving the conventional transportation engineering paradigm into a more sustainable one. Mr. Lockwood is internationally recognized for his work on context-sensitive street design, road diets, and traffic calming, winning awards from the ITE, APA, Renew America, and the CNU; but, more importantly, improving the main streets, arterial corridors, and downtowns, in which he has worked from social, economic, and environmental perspectives. Mr. Lockwood and his Glatting Jackson colleagues have helped the reform movements at several state departments of transportation through policy work, leading pilot projects, and training.

**Joel F. Mann, AICP,** is a professional planner with the community planning firm of Glatting Jackson Kercher Anglin, Inc. Mr. Mann has a Master of Regional Planning degree from the University of North Carolina at Chapel Hill. His principal interest is the development of policy and land use regulations that facilitate the development and construction of high-quality urban environments. Mr. Mann also has a strong interest in the connections between transportation infrastructure and physical planning. He works to develop plans that promote balanced multi-modal transportation, that enhance potential for land development and urban design while fitting within community context.

**Melanie McCavour** (BSc, MEnv (in progress)): holds a degree in Biology, and has taught science and organized student-led environmental initiatives over the last 10 years. She has several years of experience in environmental assessment review, both in Jalisco, Mexico (environmental effects of resort construction), and in Belize (hydro project effects on local communities).

**Jonathan Moorman** took a BA in English Literature at McGill University in 1995, and is currently enrolled in the master's program of Environmental Impact Assessment at Concordia University. He is also a professional musician (since 1991).

**Robert Moriarity** holds a Bachelors of Health Sciences from the University of Ottawa and is a Masters student in the Environmental Assessment program at Concordia University, Montreal. He has a strong interest in many fields relating to the environment. The reciprocal relationship between environment and health is his current interest.

**Munaf von Rudloff** completed a Bachelor of Arts in International Develoment Studies in 2007 and has worked in public health research, specifically HIV/AIDS in Africa. Currently enrolled in a Masters in Environmental Impact Assessment at Concordia University, Montreal.

**John Norquist**, former Mayor of Milwaukee from 1988-2004 and currently president and CEO of the Congress for the New Urbanism, received widespread recognition for championing the removal of a .8 mile stretch of elevated freeway, clearing the way for an anticipated $250 million in infill development in the heart of Milwaukee. A leader in national discussions and champion of plans to replace freeways with boulevards, Norquist is the author of *The Wealth of Cities*, and has taught courses in urban policy and urban planning at the University of Chicago, University of Wisconsin-Milwaukee School of Architecture and Urban Planning, and at Marquette University.

**Pieter Sijpkes** moved from Holland to Canada in 1966 and has taught architecture at McGill since 1976. His main interest has always been in structures—social and physical. Living in Pointe St. Charles over the last 30 years has allowed him to get involved in many social aspects of architecture. His early training in architectural engineering and his life-long interest in natural and man-made structures has been the basis for several of his courses, and has informed the way he sees the world.

**Craig Townsend**, BA UBC, MEDes (Planning) Calgary, PhD Murdoch. Assistant Professor in the Department of Geography, Planning and Environment at Concordia University in Montreal, his current research is examining the impacts of rail rapid transit systems on the built environment and social equity in Bangkok, Thailand, as well as transportation and land use policies in Canada.

**About the Editors**

**Pierre Gauthier** (BArch UdeM, MArch Laval, PhD McGill (urbanisme)) est professeur agrégé au département de géographie, urbanisme et environnement de l'Université Concordia. Ses recherches et son enseignement portent sur l'étude de la forme urbaine, l'urbanisme physico-spatial et le design urbain. Il s'intéresse de près à l'évolution du cadre bâti des quartiers centraux des villes de Québec et de Montréal, à l'histoire des pratiques de développement et d'aménagement et à l'impact des théories normatives de l'urbanisme sur la forme urbaine.

**Jochen Jaeger** received his PhD from the Department of Environmental Sciences at the Swiss Federal Institute of Technology (ETH) in Zurich, Switzerland in 2000. He is an Assistant Professor in the Department of Geography, Planning and Environment at Concordia University in Montreal. His research interests are landscape ecology, road ecology, ecological modeling, environmental impact assessment, and urban sprawl.

**Jason Prince** holds a Master's in Urban Planning from McGill University and is a member of the Ordre des urbanistes du Québec, with over 15 years experience in community economic development and the social economy. He currently coordinates a community-university research alliance (CURA), led by McGill's School of Urban Planning. Making Megaprojects Work for Communities supports action research around two contiguous mega-projects, the Turcot Interchange and the McGill University Health Centre.